Parth Chandulal Kalaria

Feasibility and Operational Limits for a 236 GHz Hollow-Cavity Gyrotron for DEMO

Karlsruher Forschungsberichte aus dem
Institut für Hochleistungsimpuls- und Mikrowellentechnik

Herausgeber: Prof. Dr.-Ing. John Jelonnek

Band 12

Feasibility and Operational Limits for a 236 GHz Hollow-Cavity Gyrotron for DEMO

by
Parth Chandulal Kalaria

KIT Scientific Publishing

Karlsruher Institut für Technologie
Institut für Hochleistungsimpuls- und Mikrowellentechnik

Feasibility and Operational Limits for a 236 GHz Hollow-Cavity Gyrotron for DEMO

Zur Erlangung des akademischen Grades eines Doktor-Ingenieurs von der Fakultät für Elektrotechnik und Informationstechnik des Karlsruher Instituts für Technologie (KIT) genehmigte Dissertation

von M.Tech. Parth Chandulal Kalaria geb. in Motimarad, Gujarat, India

Tag der mündlichen Prüfung: 29. Juni 2017
Referenten: Prof. Dr.-Ing. John Jelonnek
Prof. Dr. Machavaram V. Kartikeyan

Impressum

KIT Scientific Publishing

Karlsruher Institut für Technologie (KIT)
KIT Scientific Publishing
Straße am Forum 2
D-76131 Karlsruhe

KIT Scientific Publishing is a registered trademark of Karlsruhe Institute of Technology.
Reprint using the book cover is not allowed.

www.ksp.kit.edu

Print on Demand 2017 – Gedruckt auf FSC-zertifiziertem Papier

ISSN 2192-2764
ISBN 978-3-7315-0717-8
DOI 10.5445/KSP/1000073581

Foreword of the Editor

The global demand for electric energy is rising constantly. The developed countries continue to consume huge amounts of energy while demand is increasing in the developing countries. That demands for new sustainable energy sources. The nuclear fusion using a magnetically confined plasma is an option for a sustainable, environmental-friendly generation of electric power. In order to accomplish nuclear fusion, the particles, deuterium and tritium, must overcome the electric repulsion to get close enough for the attractive strong nuclear force to take over and fuse the particles. That requires the creation of a plasma at extremely high temperatures of above 100 million Kelvin. One possible heating and current drive system to reach and finally control those high temperatures is the Electron Cyclotron Resonance Heating and Current Drive (EC H&CD). Gyrotrons are the high-power microwave sources used to generate the RF power for EC plasma heating and current drive. Presently, ten gyrotrons, each generating an RF power of 1 MW CW at a frequency of 140 GHz, are successfully operating at Wendelstein 7-X (W7-X) stellarator at Greifswald (Germany). In future, twenty-four 170 GHz, 1 MW CW gyrotrons will be installed at the ITER tokamak at Cadarache (France). After ITER, the DEMOnstation power plant (DEMO) is planned. DEMO will show the capability to produce net electrical energy using nuclear fusion.

In Europe, the EUROfusion program coordinates the research work towards a future DEMOnstration fusion power plant. According to the EUROfusion baseline 2012, an operating frequency of up to 240 GHz is required for the EC current drive. Considering a conventional hollow-cavity gyrotron technology, the expected RF output power of a single gyrotron needs to be higher than 1 MW with an interaction efficiency of better than 35 %. Fast frequency step-tunability in steps of 2 – 3 GHz is required for frequency steering of the RF beam. Multi-purpose operation in leaps of 30 - 40 GHz is requested additionally. As the thermal loading on the cavity wall rapidly scales with the operating frequency, the maximum thermal loading of the cavity wall is determining the maximum achievable output power per tube. Very high-order transversal electric (TE) modes with eigenvalues larger than 100 need to be considered in the gyrotron design. Therefore, a prerequisite for every future fusion gyrotron design is to do a proper component's design, to do an optimization of the operating parameters and to find a suitable start-up scenario. Equally important is to push the limits for the maximum achievable RF output power.

In his work, Dr.-Ing. Parth Chandulal Kalaria presents a systematic feasibility study in combination with a detailed tolerance analysis for a possible conventional-type hollow-cavity DEMO gyrotron. He investigates the possible gyrotron performance considering realistic beam parameters and material properties in the simulation setup. The possibility of multi-purpose operation and fast frequency step-tunability are also numerically demonstrated. Dr.-Ing. Parth Chandulal Kalaria analyses the effects of an axial misalignment of the electron beam and, additionally, the effects of space-charge neutralization on single mode gyrotron operation. To verify the maximum achievable RF output power, the effects of mode competition have been systematically verified. An innovative triode-type start-up is proposed to push the limit for the maximum RF output power.

As such, Dr.-Ing. Parth Chandulal Kalaria has provided the gyrotron community with an important investigations of the principle theoretical feasibility of a hollow-cavity gyrotron for future DEMOnstration fusion power plant.

Kurzfassung

Die rasche Zunahme der Weltbevölkerung einerseits und die begrenzte Verfügbarkeit fossiler Brennstoffe andererseits deuten auf die Notwendigkeit alternativer Energiequellen hin. Eine vielversprechende Quelle für saubere, sichere und zuverlässige Energie ist die kontrollierte thermonukleare Fusion in magnetisch eingeschlossenen Plasmen. Um die technische und wirtschaftliche Machbarkeit der Energieerzeugung aus kontrollierter Kernfusion zu zeigen, wurde der Bau eines Demonstrationskraftwerks (DEMO) vorgeschlagen. In diesem werden Elektronen-Zyklotron-Resonanz-Heizung und -Stromtrieb (ECRH&CD) des Fusionsplasmas eine Schlüsselrolle spielen. Die dafür notwendige (Sub-)Millimeterwellen-Strahlung wird von Gyrotrons zur Verfügung gestellt. Zusätzlich zu hoher Ausgangsleistung, um einen ausreichenden Fusionsgewinn zu erzielen, sind Gyrotrons mit Frequenzen bis zu 240 GHz und einer Hohlraum-Wechselwirkungseffizienz von mehr als 35 % wünschenswert. Außerdem werden eine schnelle Frequenzdurchstimmbarkeit in Schritten von 2 bis 3 GHz sowie eine langsame Frequenzdurchstimmbarkeit in Sprüngen von 30 bis 40 GHz für die Kontrolle von Plasmainstabilitäten bzw. für Mehrzweckanwendungen gefordert.

Im ersten Teil der vorliegenden Arbeit wurde die Machbarkeit eines hochentwickelten DEMO-kompatiblen Hohlleiter-Gyrotrons für stabilen Dauerstrich (CW) Betrieb auf Grundlage strenger Toleranzstudien demonstriert. Eine geeignete Reihe von Betriebsmoden des Gyrotrons wurde auf Grundlage des Mehrfrequenzbetriebs und der Kriterien für Frequenzdurchstimmbarkeit ausgewählt. Unter Benutzung des vorgeschlagenen systematischen Designansatzes wurden die physikalischen Parameter des Wechselwirkungsraumes (der Cavity) so optimiert, dass die maximale Ausgangsleistung und Effizienz bei einer sinnvollen Ohmschen Wandbelastung von 2 kW/cm^2 erreicht wurden. Unter Berücksichtigung realistischer Elektronenstrahlparameter (Geschwindigkeitsverbreitung, radiale Breite) im Bereich der Cavity und einer realistischen elektrischen Leitfähigkeit des voraussichtlichen Hohlraummaterials (Glidcop) lieferte die entworfene Cavity in Simulationen eine stabile Ausgangsleistung von 920 kW. Zusätzlich zu Mehrfrequenzbetrieb bei 170 GHz, 203 GHz, 236 GHz und 269 GHz wurde die schnelle Frequenzdurchstimmbarkeit im Bereich von $\pm$10 GHz um die Mittenfrequenz von 236 GHz numerisch demonstriert. Aufgrund der geringen Wellenlänge und der hohen Betriebsmode ist das Gyrotron anfällig für Fehlausrichtungen des Elektronenstrahls. Der Einfluss axialen Versatzes auf die Gyrotron-Wechselwirkung wurde mit einem auf Makroelektronen basierenden Simulationsansatz untersucht. Für die Auslegung der Cavity von Dauerstrich-Fusionsgyrotrons sind die Effekte der Raumladungsneutralisierung entscheidend. Die Änderung der Strahlparameter aufgrund des Neutralisationsvorgangs wurde in selbstkonsistenten Mehrmodensimulationen untersucht, und das RF-Verhalten des vorgeschlagenen Hohlraums wurde für den CW-Betrieb verifiziert.

Um die Gesamtzahl der Gyrotrons pro Anlage zu reduzieren und um einen kostengünstigen und zuverlässigen Betrieb zu ermöglichen wird für DEMO eine hohe Leistung pro Röhre benötigt. Entsprechend wurden im zweiten Teil dieser Arbeit die Betriebsgrenzen des DEMO-Gyrotrons betrachtet. Die Auswirkungen des Modenwettbewerbs auf den Gyrotronbetrieb wurden gründlich untersucht und generische Ansätze wurden vorgeschlagen, um die Eigenwertgrenze für die Auswahl der Arbeitsmoden zu bestimmen. Alle diese Ansätze weisen auf eine Moden-Eigenwertgrenze von 125 für das vorgeschlagene 236 GHz Hohlleitergyrotron hin; und auf dieser Grundlage wurde ein neues 1,5 MW Gyrotron-Design für DEMO vorgeschlagen. Darüberhinaus wurden die Kontrolle des Modenwettbewerbs und die Modenauswahl unter Verwendung eines fortgeschrittenen Trioden-Startup-szenarios demonstriert.

Die diskutierten systematischen Entwurfsstudien und die gründliche Stabilitätsanalyse sind für die Entwicklung der nächsten Generation von Fusionsgyrotrons mit verbesserter Ausgangsleistung und robustem Betrieb von Bedeutung.

Abstract

The rapid increase in the world population and limited availability of fossil fuel suggests the necessity for alternative energy sources. The controlled thermonuclear fusion in magnetically confined plasma is a promising option for clean, safe and reliable energy source. A DEMOnstration power plant (DEMO) is proposed to be built to prove the technical and economical feasibility of controlled thermonuclear fusion energy. Electron Cyclotron Resonance Heating and Current Drive (ECRH&CD) of the fusion plasma plays a key role in such a DEMO and the necessary mm-wave/sub-millimeter wave radiation is provided by gyrotrons. Along with high output power, to achieve sufficient fusion gain, it is desirable to use gyrotrons with frequencies up to 240 GHz and cavity interaction efficiency higher than 35 %. In addition to this, fast frequency tunability in steps of 2 - 3 GHz and slow frequency tunability in steps of 30 - 40 GHz is also requested for plasma instability control and multi-purpose applications, respectively.

In the first part of this work, the feasibility of an advanced DEMO-compatible hollow-cavity gyrotron was demonstrated, along with rigorous tolerance studies for stable Continuous Wave (CW) operation. The suitable operating mode series has been selected according to the multi-frequency gyrotron operation and frequency-tunability criteria. Using a suggested systematic cavity design approach, the physical parameters of the cavity were optimized such that the maximum output power and efficiency were achieved with the reasonable Ohmic cavity wall-loading of 2 kW/cm^2. Considering realistic electron beam parameters in the cavity (velocity spread, radial width) and a realistic conductivity of the anticipated cavity material (Glidcop), the designed cavity yielded stable output power of 920 kW (according to simulation). The fast-frequency step-tunability was numerically demonstrated within the frequency range of ±10 GHz around the center frequency of 236 GHz, along with the multi-frequency operation at 170 GHz, 203 GHz, 236 GHz and 269 GHz. Because of the small wavelength and high order operating mode, the gyrotron is susceptible to beam misalignment. The influence of axial-misalignment on gyrotron interaction was studied using a macro-electron based simulation approach. For CW fusion gyrotrons, the effects of space-charge neutralization is critical for the hollow-cavity design. The variation of beam parameters due to the neutralization process was considered in multi-mode, self-consistent simulations, and the RF behavior of the proposed cavity was verified for CW operation.

To reduce the overall number of tubes per plant and to facilitate cost-effective and reliable operation, high power per tube is requested for DEMO. The operational limits of the DEMO gyrotron were investigated in the second part of this work. The effects of mode competition on gyrotron operation were rigorously studied and generic approaches were suggested to determine the mode eigenvalue limit for operating mode selection. All these methods specified a mode eigenvalue limit of 125 for the proposed 236 GHz hollow-cavity gyrotron and based on it, a new 1.5 MW gyrotron design was proposed for DEMO. Furthermore, the mode competition control and mode selectivity were demonstrated using an advanced triode-type start-up scenario.

The discussed systematic design studies and exhaustive stability analysis are significant for the development of next generation of fusion gyrotrons with improved output power and robust operation.

Contents

Used symbols, variables and abbreviations

Symbols and variables

b	magnetic field compression
B	magnetic field
B_0	magnetic field at cavity center
$B(r)$	magnetic field inside plasma
D	value of axial beam misalignment
e	elementary charge
E	electric flux density
f	operating frequency
f_{B}	basic frequency of RF window
f_{c}	relativistic electron cyclotron frequency
f_{cut}	cutoff frequency
$Ge_{\mathrm{m,n}}$	beam wave coupling coefficient of mode $\mathrm{TE}_{\mathrm{m,n}}$
h	rms height of rough surface profile
I_{b}	beam current in ampere
k	wave propagation constant
$k_{\perp}, k_{\parallel}$	perpendicular and parallel wavenumber
m_e	electron rest mass
p	momentum
$p_{\perp}$	transverse electron momentum

P_{beam}	total power of electron beam
P_{c}	recovered power by collector
P_{out}	output power of RF beam
P_{Ω}	power due to ohmic losses
$Q_{\mathrm{diff}}, Q_{\Omega}$	diffractive quality factor and ohmic quality factor
r_{b}	average beam radius at cavity center
r_{c}	caustic radius
r_{cav}	radius of cavity interaction section
r_{E}	radius of emitter center
r_{gc}	average guiding center radius
r_{L}	Larmor radius
r_{w}	radius of waveguide
s	number of harmonics
v	velocity in m/s
$v_{\perp}, v_{\parallel}$	perpendicular and parallel velocity component
V_{b}	beam energy in keV
V_{c}	cathode voltage in kV
W_{kin}	energy of electron beam
Z_0	free space impedance (376.73 Ω)

Greek variables

α	velocity ratio ($\alpha = \beta_{\perp}/\beta_{\parallel}$) or pitch factor
β	normalized velocity ($\beta = v/c$)
$\beta_{\perp}, \beta_{\parallel}$	perpendicular and parallel normalized velocity
γ	Lorentz factor
γ_{eff}	relativistic factor before space-charge neutralization
δ	skin depth of conducting material
$\delta\alpha$	spread in velocity ratio
$\delta\beta_{\perp}$	spread in perpendicular velocity

Δ	normalized detuning factor
Δr_{gc}	electron guiding center radius
ΔV_{hollow}	depression voltage in hollow cavity gyrotron
ε_0	free space permittivity
ε_r	relative permittivity of material
η	interaction efficiency
η_c	collector efficiency
θ_B	Brillouin angle
λ	wavelength
μ_0	permeability of free space
ρ_Ω	cavity wall-loading
σ	conductivity of material
$\chi_{m,n}$	eigenvalue of the cylindrical waveguide $TE_{m,n}$ mode
ω_{RF}	frequency of RF oscillation
Ω_c	relativistic angular electron cyclotron frequency

Abbreviations

ACI	after cavity interaction
CD	current drive
CW	continuous wave
CPR	controlled porosity reservoir
DEMO	demonstration power plant
EC	electron cyclotron
ECCD	electron cyclotron current drive
ECM	electron cyclotron maser
ECRH	electron cyclotron resonance heating
ECRH&CD	electron cyclotron heating and current drive
EU	European Union

FGMC	fundamental Gaussian mode content
GO	geometric optics
IFERC	International Fusion Energy Research Centre
IHM	Institute for Pulsed Power and Microwave Technology
IMIG	inverse magnetron injection gun
KIT	Karlsruhe Institute of Technology
LP	long pulse
MIG	magnetron injection gun
MOU	matching optical unit
NBI	neutral beam injection
NTM	neoclassical tearing mode
PIC	particle in cell
QOL	quasi optical launcher
RF	radio frequency
SIM	surface impedance model
Self – C	self-consistent
SPC	Swiss Plasma Center
TE	transverse electric
TM	transverse magnetic
VDEs	vacuum electron devices
W7 – X	wendelstein 7-x

1 Introduction

In this chapter, the applications of high power gyrotrons to thermonuclear fusion plasma experiments are discussed, followed by the description of gyrotron requirements for the next generation DEMOnstration power plant (DEMO). The overview of gyrotron principles and its main components are also presented. The scope of this work is discussed in the last section of this chapter.

1.1 Fundamentals of fusion plasma experiments

By the year 2050, the amount of energy requirement is expected to be doubled because of the rapid increase in the world population [IAE00]. Around 80 % of the present energy requirement are being satisfied by the fossil fuels, but they are limited in nature and generate unacceptable greenhouse gases [Fus12]. The renewable energy sources like solar and wind are providing green energy, but their capabilities to fulfill next generation energy demand are under discussion [Che11]. Alternatively, the controlled thermonuclear fusion in magnetically confined plasma represents one of the most promising long-term and clean energy supplies for the future.

As per the Lawson-criterium [Law57], low energy fusion reaction is only possible with the lightest nuclei. The hydrogen isotopes namely Deuterium

($^2_1$D) and Tritium ($^3_1$T) are used in the most effective fusion reaction. The complete process can be represented as,

$$^2_1\mathrm{D} + ^3_1\mathrm{T} \rightarrow\ ^4_2\mathrm{He} + n \tag{1.1}$$

The energy released in this fusion reaction is based on Einstein's famous equation $E = \Delta mc^2$ [Ein35]. The total energy of 17.6 MeV has been generated per reaction, which is divided among the neutron (14.1 MeV) and helium atom (3.5 MeV). As both the hydrogen isotopes are positively charged, therefore, for a successful fusion reaction, the kinetic energy of around 10 keV ($\sim$ 116 Mio. K) is required to overcome the ion-ion repulsive force called coulomb barrier.

Since, it is not possible to generate such high temperature using a single heating system, thus, more than one heating system is used simultaneously to reach the desired temperature. The main heating systems are:

- Ohmic heating: Here, the conducting plasma behaves as a secondary winding of a transformer and the plasma current is induced by slowly increasing the current through an electromagnetic winding, which is linked to the plasma torus. The generated heat depends on the plasma resistance and the induced current. With the increase in temperature, the Ohmic heating becomes less effective because of the reduction of the plasma resistance.

- Neutral Beam Injection (NBI): High energy neutral atoms are generated outside the torus. This neutral beam passes the high magnetic field region and thereafter, reaches the plasma region, where it gets ionized. The generated ions are trapped in the strong magnetic field and transfers their energy to the plasma ions and electrons by repeated collisions [HDG+09].

- Radio-Frequency (RF) heating: According to the Lorentz force, the charged particles in the plasma (electrons and ions) rotate around the static magnetic field with their cyclotron frequency. Based on the cyclotron interaction, RF energy is transferred to the charged particle in the form of thermal kinetic energy. According to the selected charge particles, this method is divided into two sub-categories namely: (a) Ion Cyclotron Resonance Heating (ICRH) [SJB+98] and (b) Electron Cyclotron Resonance Heating (ECRH) [EKP+12]. The electromagnetic waves for the ICRH systems are generated using tetrode based sources, while high-power gyrotrons are used as microwave source for the ECRH application [JAF+14].

Compared to the other plasma heating systems, the advantages of the ECRH system includes localized power deposition, small port size, controlled absorption and possibility of remote steering [GLP+11]. It also supports quasi-optical propagation of the beam, which simplifies the launching system. Hence, a launcher can be placed far away from the hot plasma surface. For both the possible configurations of fusion plasma experiments (e.g. tokamak and stellarators), the ECRH systems are proved to be effective and are also planned for future plasma experiments [PTZ+12].

High-power RF sources operating in a frequency range between 100 GHz and 300 GHz are required for ECRH in fusion plasma experiments [Sch15]. Gyrotron oscillators (gyrotrons) are the only sub-millimeter and millimeter wave sources, capable to produce high-power (few kW to several MW) electromagnetic (EM) waves [Thu16]. The main applications of high-power ($\sim$ 1 MW), high-frequency (100 GHz – 300 GHz) gyrotrons are plasma start-up, Electron Cyclotron Resonance Heating (ECRH), non-inductive Current Drive (CD), Collective Thomson Scattering (CTS) diagnostics and plasma stability control in plasma experiments relevant for nuclear fusion research. As an example, for the Wendelstein 7 - X (W7-X) stellarator in

Greifswald (Germany), in total ten 140 GHz, 1 MW CW (1800 s) hollow-cavity gyrotrons are completely installed and are being operated successfully to fulfill the total ECRH requirement [EBG+14]. The RF-power is transmitted using quasi-optical mirror based transmission systems and is launched into the plasma vessel with the help of four equatorial ports [EBL+05]. For the ITER tokamak in Cadarache (France), 24 Continuous Wave (CW) ($\sim$ 3600 s) gyrotrons with an operating frequency of 170 GHz are planned for a total power requirement of 20 MW [JAA+15] [KST+08]. The major goal of ITER is to achieve the record fusion power gain ($Q = P_{fusion}/P_{heating}$) of 10 and to validate possibilities of steady-state controlled plasma operation.

1.2 DEMO: the next step towards fusion power

After ITER, the first prototypes of fusion power plant are foreseen, termed DEMOnstration power plants (DEMO) [LST11]. DEMO is notably intended to be the first fusion reactor to generate electrical power. In addition to net energy generation, the DEMO plant is aiming to prove tritium self-sufficiency, adequate reliability, and to demonstrate all the technological constructions of commercial power plants [WNT95], [HOA+05]. As DEMO is the last step before commercial Fusion Power-Plants (FPP), it is also important to resolve the exciting physical challenges associated with steady-state fusion operations [FBB+16]. DEMO must be designed to facilitate remote handling capabilities with fast maintenance schemes to ensure a requisite availability. Additionally, in order to support development of FPP, DEMO is also targeted to minimize electricity cost from fusion [Rom12].

Conceptual design studies of DEMO have been initiated by many research groups around the world, and various approaches/designs are proposed, for example, the European Union-DEMO (EU-DEMO) [FGL+13] [MCC+07]

[FKW+14], the Korean-DEMO (K-DEMO) [KIP+16] [KKO+13], the China Fusion Engineering Test Reactor (CFETR) [FZZ+09] [TWL+15, WDQ+13], the Japanese DEMO [PBB+09] [STK+06a] [SBF+12] [STK+06b] etc.. For all the above stated DEMO designs, Electron Cyclotron Resonance Heating and Current Drive (ECRH&CD) is selected as a fundamental plasma heating system and high power gyrotrons with frequency more than 200 GHz are requested to fulfill plasma heating and current drive requirements. The detailed specification of the ECRH&CD systems of EU-DEMO is discussed in [PTZ+12], [STK+06c], [FKW+14] [WAA+15]. Based on these specifications, the requirements of the DEMO gyrotrons are discussed in the next section.

1.2.1 Requirements for future DEMO gyrotrons and motivation for this work

As per the EUROFusion baseline of 2012, the maximum ECRH power of 50 MW is required at various operation stages of DEMO [GBB+15]. Unlike ITER, considering steady-state operation of the machine, the system efficiency is optimized to improve Electron Cyclotron Current Drive (ECCD) performance, instead of plasma heating. To select optimum ECCD frequency, detailed analyses were performed in [PTZ+12]. For steady-state DEMO operation, the operating frequencies range of around 230 GHz were suggested for efficient ECCD operation using an upper torus port. The various operational stages of a DEMO plant (e. g. plasma start-up, non-inductive current drive, and bulk heating) require different microwave frequencies [TFK+15]. It is also beneficial to use the same gyrotron as a multi-purpose tube in a fusion facility. Furthermore, the same operational stage in facilities with different torus aspect ratios (torus radius/plasma radius) would require different microwave frequencies [Mor92]. The requested multi-purpose requirements can be fulfilled by a proper "multi-frequency"

Table 1.1: Design goals of DEMO gyrotrons.

Goals	
Frequency	230 – 240 GHz
Output power	$\sim 1 - 2$ MW
Total gyrotron ("plug-in") efficiency	> 60 %
Beam-wave interaction efficiency	> 35 %
Frequency step for fast tunability	2 – 3 GHz
Frequency step (slow) for multi-frequency operation	30 – 40 GHz

gyrotron design. Considering ITER gyrotron operation at 170 GHz and the possibilities of multi-frequency operation, a frequency around 236 GHz is selected for DEMO. The selection of operating frequency is detailed in section 2.1.1. The minimum expected output power from hollow-cavity gyrotrons is 1 MW. This output power requirement can also be increased up to 2 MW by using more complex, coaxial cavity gyrotrons [FIK+14]. In addition to the high-output power requirements, in order to maintain high energy gain of the DEMO plant, a gyrotron with a total efficiency of more than 60 % is requested. This can be achieved by using a multi-stage depressed collector. This leads to the requirement of a minimum cavity interaction efficiency of 35 %.

Additionally, for plasma stability control using a simple fixed launcher, fast frequency step-tunability (in few seconds) is needed in steps of 2 - 3 GHz [WAA+15]. It could be noted that multi-propose, multi-frequency operation is possible with the standard single disk windows, however, the operation with fast-frequency step-tunability can be only achieved with the application of Brewster-type broadband windows [GSA+14]. As per the European Union (EU) 2012 baseline of DEMO (aspect ratio of 4.0), the detailed design goals for gyrotron development are summarized in Table 1.1.

During the last decade, advanced technologies are being developed by the various research groups to demonstrate gyrotron operation with ITER specifications (170 GHz, 3600 s operation). Following the successful development of 140 GHz, 1 MW gyrotrons for the W7-X stellarator, 170 GHz coaxial-cavity and hollow-cavity ITER gyrotrons are under investigation at the Institute for Pulsed Power and Microwave Technology (IHM) of the Karlsruhe Institute of Technology (KIT) along with the partner institutes. The highest output power of 2.2 MW has been achieved with the 170 GHz coaxial-cavity short-pulse, pre-prototype tube with the total efficiency of 46 % (using a single stage-depressed collector)[RPK+10] [RAA+17]. The record efficiency and energy operation was achieved by the Japan 170 GHz ITER hollow-cavity gyrotron with 57 % of total efficiency at 0.8 MW output power (pulse length: 3600 s) [SKK+09]. Prototype of ITER gyrotrons were also developed at the Institute of Applied Physics, Russia and long-pulse gyrotron operation (1000 s) was verified with an output power of 1 MW [LDM+12]. Recently in Japan, the successful gyrotron operation at the operating frequency of 203 GHz has been demonstrated with output power of 0.42 MW (5 s) and interaction efficiency of 25.1 % [IOK+16].

The DEMO ECRH&CD system demands highly efficient gyrotrons (interaction efficiency > 35 %) at an operating frequency of more than 230 GHz. At this frequency range, the operation of MW-class fusion gyrotrons was not previously verified, and a systematic feasibility study is requested. Considering the EUROfusion baseline of 2012, for the first time, the conceptual design of DEMO gyrotrons has been initiated at IHM-KIT. Compared to the hollow-cavity gyrotron, higher output power per tube can be achieved using the coaxial-cavity gyrotron, but its capabilities of long-pulse operation are not yet proven. Additionally, the complexity of the design and the stability of operation are under discussion for coaxial-cavity gyrotrons. Thus, to achieve long-pulse MW-level gyrotron operation with a simple and robust design, the hollow-cavity gyrotron is selected in this work.

Considering the present challenges and technical constraints of high-frequency, high power hollow-cavity gyrotrons, the feasibility of the advanced multi-frequency, frequency step-tunable DEMO gyrotron shall be systematically investigated, along with its performance limit analysis.

1.3 Gyrotron components and their principles

Gyrotrons are fast-wave, Vacuum Electron Devices (VDEs), which generate high power electromagnetic waves in the sub-millimeter/millimeter frequency range, by following the principles of Electron Cyclotron Masers (ECM) instability [Chu04]. The 2-D sketch of a high power hollow-cavity gyrotron is presented in Figure 1.1. In the presence of magnetic and electric fields, an annular electron beam is generated by the Magnetron Injection Gun (MIG). The generated electron beam is focused towards the interaction cavity with the help of a very high magnetic field produced by superconducting magnets. Inside the cavity, the electron beam interacts with and transfers a part of its kinetic energy to a transverse electric field (TE mode). Using an quasi-optical converter and metallic mirrors, the electromagnetic wave is separated from the electron beam and is transmitted through an RF window. The spent electron beam is absorbed by the collector. Using either a single-stage or a multi-stage depressed collector, the energy of the spent electron beam is further recovered to improve the overall tube efficiency.

1.3.1 Basic principle of gyrotrons

The gyrotron interaction phenomenon is also termed as electron cyclotron interaction. The detailed description of the gyrotron interaction can be found in [FDJ^{+}99] [SJ81] [FGPY77] [KBT10]. In the presence of a strong magnetic field produced by superconducting magnets, the electron energy can

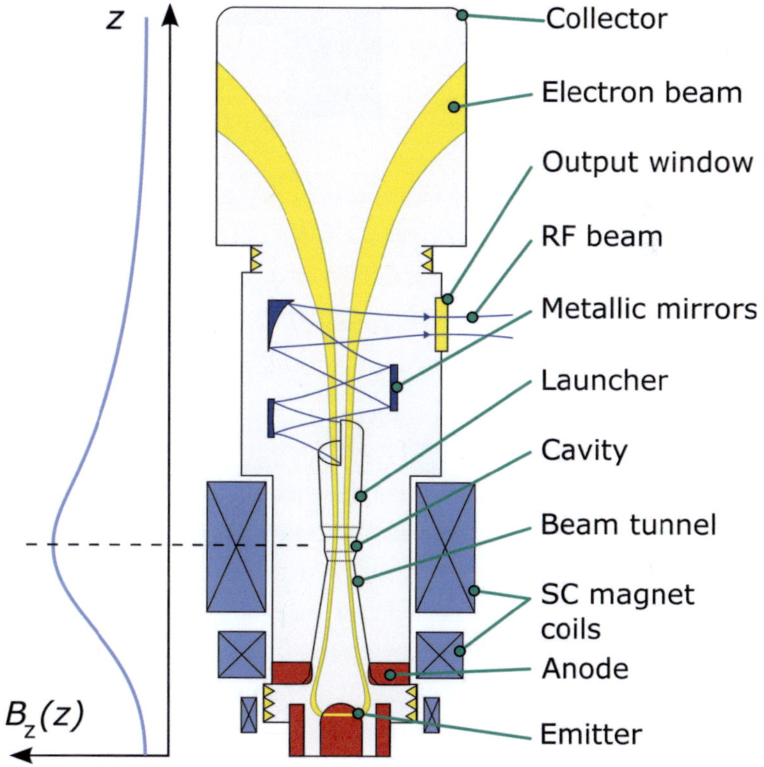

Figure 1.1: 2D-sketch of a high power fusion gyrotron with radial output. The magnetic field along the axis is also shown in the left side of the figure.

be described using the weakly relativistic factor. In such cases, the effective relativistic electron angular cyclotron frequency is given by,

$$\Omega_c = 2\pi f_c = \frac{eB}{m_e \gamma}, \tag{1.2}$$

where, f_c is the relativistic electron cyclotron frequency, B is the magnetic field, m_e is the electron rest mass and the relativistic factor γ can be described by,

$$\gamma = \frac{1}{\sqrt{1-(v/c)^2}} = 1 + \frac{W_{kin}}{m_e c^2} \approx 1 + \frac{W_{kin}}{511}, \tag{1.3}$$

where, W_{kin} is the energy of electron beam in keV and v represents the velocity of electron. For particular beam energy, the magnetic field strength defines the radiation frequency (see equation 1.2). The distance between actual spiral trajectory and electron guiding center is known as the Larmor radius (r_L), which is given by,

$$r_L = \frac{\gamma m v_\perp}{eB}, \tag{1.4}$$

where, $v_\perp$ is the perpendicular velocity component. In the presence of RF electric field, according to the electron relative phase position, they are accelerated or decelerated and the value of the Lorentz factor changes accordingly. In Figure 1.2, the interaction of the gyrating electron and electric field is presented with two examples. When the electron is in the accelerating phase (top position in Figure 1.2) it gains energy, and as per equation 1.3, the value of γ increases due to increase in its velocity. Because of the increased velocity, electrons leave this undesired phase position quickly. This change in γ also leads to the reduction in the cyclotron frequency (Ω_c) with increased Larmor radius (r_L).

In the deceleration phase (bottom position in Figure 1.2), the relativistic factor (γ) decreases due to the reduction in the electron energy and the cyclotron frequency (Ω_c) increases. When the cyclotron frequency (Ω_c) is selected somewhat lower than the angular frequency of the RF waves (ω_{RF}), the difference in both the frequencies ($\omega_{RF} - \Omega_c$) decreases in the deceleration phase. This condition reduces the relative phase difference between gyrating electrons and the EM wave. The electrons stay longer in the matching phase position where, the electrons energy is extracted. In summery, when the electron cyclotron frequency (Ω_c) is selected marginally lower than the wave frequency, the particles continuously accumulate at relatively same phase position, where they transfer energy to the EM wave, which is termed as "bunching". For the successful energy transfer from electron beam to EM wave, the frequency mismatch ($\omega_{RF} - \Omega_c$) should be selected small but positive.

The difference in wave frequency and cyclotron frequency is represented by the detuning factor. The normalized detuning parameter can be calculated as [Bra83],

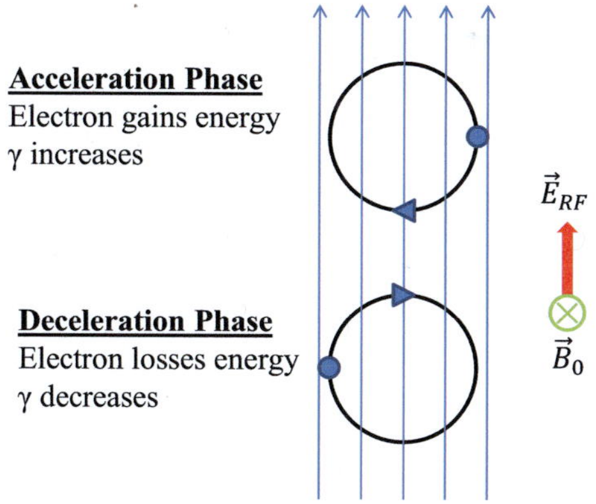

Figure 1.2: Principle of electron cyclotron interaction.

$$\Delta = \frac{2}{\beta_{\perp}^{2}} \left(\frac{\omega_{\mathrm{RF}} - s \cdot \Omega_{c}}{\omega_{\mathrm{RF}}} \right) = \frac{2}{\beta_{\perp}^{2}} \left(\frac{\omega_{\mathrm{RF}} - s \cdot \left(\frac{e \cdot B_{0}}{\gamma \cdot m_{e}} \right)}{\omega_{\mathrm{RF}}} \right). \tag{1.5}$$

Here, $\beta_{\perp} = v_{\perp}/c$ is the normalized transverse velocity of electrons and s represents the number of harmonics. For fundamental mode operation, the value of s is 1. The electron passes through the cavity region with an axial velocity of $v_{\parallel}$. With the axial motion of electrons, the resonance condition modifies to,

$$\omega_{\mathrm{RF}} - k_{\parallel} v_{\parallel} \geq \Omega_{\mathrm{c}}, \tag{1.6}$$

where, the term $k_{\parallel} v_{\parallel}$ is know as the Doppler term. The total wave number is given by,

$$k = \frac{2\pi}{\lambda} = \frac{2\pi f}{c}, \tag{1.7}$$

$$k^{2} = k_{\parallel}^{2} + k_{\perp}^{2}, \tag{1.8}$$

where, $k_{\parallel}$ and $k_{\perp}$ is the parallel and perpendicular wavenumber, respectively. In the case of high harmonics gyrotrons, the equation 1.6 is updated to

$$\omega_{\mathrm{RF}} - k_{\parallel} v_{\parallel} \geq s\Omega_{\mathrm{c}}. \tag{1.9}$$

From equation 1.6, it is clear that changes in electron energy not only change the cyclotron frequency, but also affect the axial velocity and correspondingly the Doppler term. In the presence of an EM wave with the phase velocity of $v_{ph} = w_{RF}/k_{\parallel}$, the changes compensate each other. Once the resonance condition is satisfied, then it maintained through out the operation. This effect is known as autoresonace [Nus04].

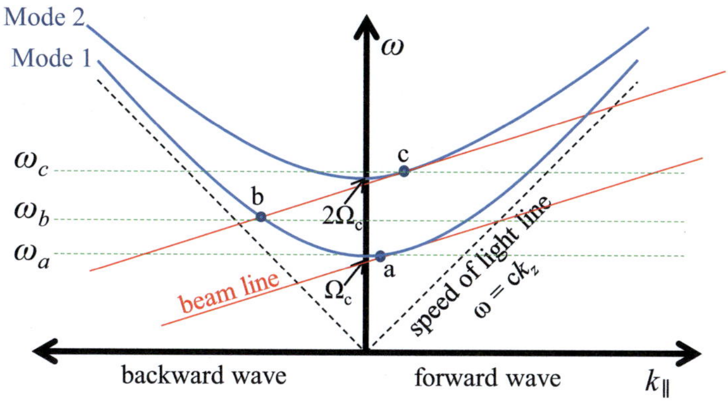

Figure 1.3: Dispersion diagram of possible operating modes with beam line and speed of light line.

The resonance condition between waveguide mode and electron beam can be illustrated using the Brillouin diagram [Zha86] [ZSS07] [Edg80] [CS81]. This diagram is also named as Dispersion diagram, in which the TE modes are represented as per equation 1.8 and the beam line follows,

$$\omega_{\mathrm{RF}} - k_{\|} v_{\|} = s\Omega_{\mathrm{c}}. \tag{1.10}$$

In Figure 1.3, the hyperbola represents the dispersion curves of two possible TE modes, while the red lines represent the beam lines. The intersection point of these two curves satisfies the resonance condition between beam and wave inside the gyrotron cavity. Point "a" indicates the gyrotron interaction point with small value of $k_{\|}$ and, here the operating frequency is close to the cut-off frequency. The possibilities of backward wave propagation is demonstrated by point "b", for which the RF frequency is higher than the cutoff frequency. The point "c" satisfies the forward wave resonance condition with Mode 2 and second harmonic of the cyclotron frequency.

1.3.2 Electron emission and beam optics

The electron beam optics system includes the Magnetron Injection Gun (MIG), beam tunnel and a super conducting magnet system. For efficient interaction, a high power, hollow-electron beam is prerequisite with sufficient transverse velocity ($v_{\perp}$). The beam generated by the MIG passes through the beam tunnel section, which is placed between gun and cavity to suppress unwanted low-frequency parasitic oscillations and avoid the backward wave propagation [Edg93].

The structure of MIGs can be classified in two types (a) diode-type gun and (b) triode-type gun. The simple schematic of the gun types are represented in Figure 1.4. Unlike the diode-type gun, in case of a triode type gun, the additional control over the beam parameters can be achieved by tuning modulating anode voltage (V_{M}). For high power gyrotrons, impregnated dispenser emitters are used, which are operated in the temperature limited region [Thu16]. The beam parameters are sensitive to the surface roughness of the emitter and emission inhomogeneity [Zha16].

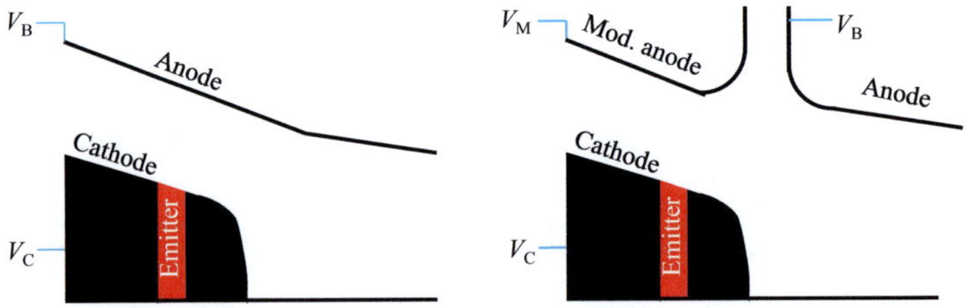

(a) (b)

Figure 1.4: Configuration of (a) diode-type and (b) triode-type Magnetron Injection Gun (MIG). Here, V_{C} and V_{B} are the cathode and anode voltage, respectively. In the case of triode start-up, the modulating anode voltage is represented as V_{M}.

The velocity ratio (α) is the ratio of the transverse velocity component to the axial velocity component,

$$\alpha = \frac{\beta_\perp}{\beta_\parallel}. \tag{1.11}$$

Where, $\beta_\parallel = v_\parallel / c$ is the normalized axial electron velocity. In the ideal case, an electron beam without velocity spread is desired for efficient gyrotron operation, but considering the practical limitations of a gun design discussed in [Tsi01], the spread in electron beam velocities is unavoidable. In the case of an adiabatic approximation, the perpendicular velocity spread is constant along the beam. From the perpendicular velocity spread, the spread in the velocity ratio can be estimated using,

$$\delta\alpha = (1+\alpha_0^2)\cdot\delta\beta_\perp. \tag{1.12}$$

Here, α_0 is the mean value of the velocity ratio, $\delta\alpha$ and $\delta\beta_\perp$ are the rms values of the spread in velocity ratio and perpendicular velocity, respectively. The effects of transverse velocity spread on gyrotron interaction is studied with greater details in section 2.3.2. The compression ratio (b) of the magnetic field at the cavity center (B_0) and the emitter surface (B_E) is given by,

$$b = \frac{B_0}{B_\mathrm{E}} = \left(\frac{r_\mathrm{E}}{r_\mathrm{b}}\right)^2, \tag{1.13}$$

and, using it, the electron guiding spread at cavity center can be determined by,

$$\Delta r_{gc} = \frac{r_\mathrm{Emax} - r_\mathrm{Emin}}{b^{1/2}}. \tag{1.14}$$

Here, r_E and r_b are the radius of the electron beam at the emitter and cavity center, respectively. In the case of slow magnetic compression along the gyrotron axis, the movement of the electrons in the magnetic field of a gyrotron can be defined using adiabatic approximation following from Busch's theorem [Che74],

$$\frac{p_{\perp}^{2}}{B(z)} = \text{const}, \tag{1.15}$$

where, $p_{\perp} = m_0 \gamma v_{\perp}$ is the transverse momentum of the electrons. It can be influenced by the design of the gun, and the choice of the operating parameters. Variation of the transverse momentum and the normalized transverse velocity along the gyrotron can be estimated using [Cho14],

$$p_{\perp}(b) = p_{\perp}(a)\left[\frac{B(b)}{B(a)}\right]^{1/2}, \tag{1.16}$$

$$\beta_{\perp}(b) = \beta_{\perp}(a)\frac{\gamma(a)}{\gamma(b)}\left[\frac{B(b)}{B(a)}\right]^{1/2}, \tag{1.17}$$

here, a and b is the position along the gyrotron axis. The normalized axial velocity competent along the gyrotron axis can be calculated as, [Pio90],

$$\beta_{\parallel}(b) = \sqrt{\beta^{2}(b) - \beta_{\perp}{}^{2}(a) B(b)/B(a)}. \tag{1.18}$$

With magnetic field compression (increase in magnetic field), the transverse momentum ($p_{\perp}$) and the normalized transverse velocity ($\beta_{\perp}$) increases, which decreases the axial momentum ($p_{\parallel}$) and the normalized parallel velocity ($\beta_{\parallel}$). This also leads to an increase of the velocity ratio (α) along the magnetic field (B).

1.3.3 Interaction section

The gyrotron interaction section or gyrotron cavity is a straight cylindrical resonator with a down-taper at its input and an up-taper at its output, in which the energetic electron beam transfers its perpendicular kinetic energy

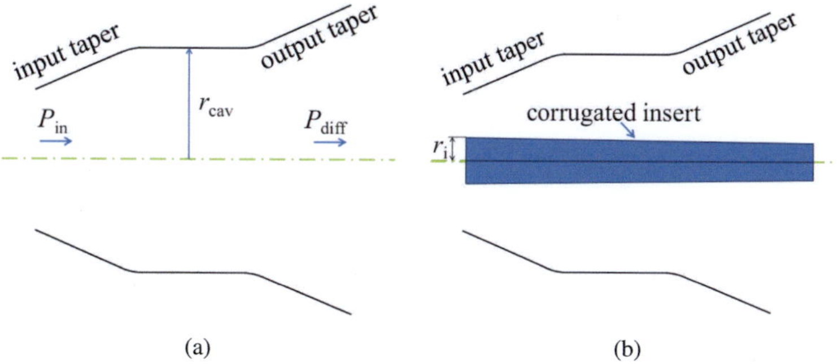

Figure 1.5: Simple schematic of (a) hollow-cavity and (b) coaxial-cavity. The interaction section radius is denoted as r_{cav} and the insert radius is r_i. The total input power and radiated power is denoted as P_{in} and P_{diff}, respectively.

to the RF wave. The cavity is placed at the position of maximum magnetic field. The physical parameters of the cavity are selected to have efficient gyrotron interaction with controlled mode competition.

The maximum cavity wall-loading is limited by the available cavity cooling system. Two different structures for the interaction section are used for high-power fusion gyrotrons namely (a) hollow-cavity and (b) coaxial-cavity. The hollow cavity is a three section cylindrical structure with input taper, straight interaction section and up-taper section (see Figure 1.5), whereas in the coaxial cavity design, an additional corrugated insert is placed at the cavity center to control mode competition. The design considerations and the selection criteria of each physical parameters of the hollow cavity structure is discussed in section 2.2.

As discussed in section 1.3.1, for successful energy transfer between the beam and the wave, the electrons are decelerated by the electric field. Over the interaction length, the TE mode remains phase-synchronized with the gyrating electron beam and supports efficient interaction. Since in gyrotrons, the interaction of an electron beam happens with a TE mode, which is very close to cut-off. Thus, the transverse wave number in the cavity is given as,

$$k_{\perp} = \frac{\chi_{\mathrm{m,n}}}{r_{\mathrm{cav}}}, \tag{1.19}$$

where, r_{cav} is the radius of the straight interaction section and $\chi_{\mathrm{m,n}}$ is the mode eigenvalue. For TE modes, $\chi_{\mathrm{m,n}}$ is the nth root of the derivative of the Bessel function (J'_{m}), which is also known as the eigenvalue of the mode TE$_{m,n}$. Here, m and n are the azimuthal and the radial index of the mode, respectively. In this work, the modes co-rotating with the beam electrons are represented with a negative sign, while counter-rotating modes are presented with a positive sign. Using equation 1.19, the cut-off frequency of the operating mode is described as,

$$f_{cut} = \frac{c\chi_{m,n}}{2\pi r_{\mathrm{cav}}}. \tag{1.20}$$

Considering the dependency of the azimuthal electrical component (E_{ϕ}) over the radial direction, the beam-wave coupling is characterized by the coupling co-efficient [Ker96],

$$Ge_{\mathrm{m,n}} = \frac{J_{\mathrm{m}\mp 1}(k_{\perp \mathrm{m,n}} r_{\mathrm{b}})}{J_{\mathrm{m}}(\chi_{\mathrm{m,n}})\sqrt{\pi(\chi_{\mathrm{m,n}}^{2} - m^{2})}}, \tag{1.21}$$

and the optimum electron beam radius for the highest mode coupling can be calculated using,

$$r_{\mathrm{b}} = \left(\frac{r_{\mathrm{cav}}\chi_{\mathrm{m}\mp 1,1}}{\chi_{\mathrm{m,n}}}\right). \tag{1.22}$$

Cavity wall loading

For continuous wave (CW) operation of high-power gyrotrons, the cavity wall-loading is critical and greatly influences the selection of the operating parameters and the cavity mode. In case of hollow cavity design, the total RF power (P_{total}) consists of the radiated power (P_{diff}) and the power loss due to the Ohmic wall-loading (P_{Ω}). Considering the total stored energy W and a uniform straight interaction section, the diffractive quality factor is given by,

$$Q_{\text{diff}} = \omega_{RF}\frac{W}{P_{\text{diff}}} = 4\pi\frac{L_{\text{R}}^2}{\lambda_0^2}\left(\frac{1}{1-\rho}\right), \tag{1.23}$$

and the Ohmic quality factor is defined by [Edg93]:

$$Q_{\Omega} = \omega_{RF}\frac{W}{P_{\Omega}} = \frac{r_{\text{cav}}}{\delta}\left(1-\frac{m^2}{\chi_{\text{m,n}}^2}\right), \tag{1.24}$$

where, ρ is the reflection from the cavity output end, L_{R} is the effective length of cavity and $\delta = 1 \ / \ \sqrt{\pi f \mu_0 \sigma}$ is the skin depth. To calculate the diffractive quality factor, a total reflection is considered from the input port and the frequency of the operating mode is close to the cut-off frequency of the cylindrical waveguide. Based on these two quality factors, the total quality factor Q is calculated as,

$$\frac{1}{Q} = \frac{1}{Q_{\text{diff}}} + \frac{1}{Q_{\Omega}} \tag{1.25}$$

Using equation 1.23 and 1.24 the Ohmic losses can be calculated as [Edg93],

$$P_{\Omega} = 2c\left(\frac{\pi f}{\mu_0\sigma}\right)^{1/2} Q_{\text{diff}} P_{\text{out}} \frac{\chi_{\text{m,n}}}{1-m^2/\chi_{\text{m,n}}^2}, \tag{1.26}$$

and selecting total surface area of cavity (S), the cavity wall-loading is given by [VZO+69],

$$\rho_{\Omega} = \frac{P_{\Omega}}{S} = \frac{Q_{\text{diff}}\cdot P_{\text{diff}}}{Q_{\Omega}\cdot S} \approx \frac{2\sqrt{\pi}}{c^3\sqrt{\mu_0\sigma}}\frac{f^{2.5}Q_{\text{diff}}}{L_{\text{R}}/\lambda\,(\chi_{\text{m,n}}^2-m^2)}P_{\text{diff}}. \tag{1.27}$$

From present capabilities of cavity cooling technologies, the maximum cavity wall-loading of 2 kW/cm^2 is selected for the DEMO gyrotron design. From equation 1.27, it is clear that the cavity wall loading is highly dependent on the operating frequency, which limits the maximum possible output power per tube for high frequency gyrotrons. This challenge is addressed with great details in the (chapter 5) of this work.

Mode competition in high-power gyrotrons

Due to the limitations in wall-loading at high operating frequencies ($f >$ 100 GHz), high-power gyrotron operation is only possible using very high order modes, which allows operation with increased cross-section. Subsequently, a fusion gyrotron operates in a high-order operating mode with dense mode spectra, which may lead to the excitation of parasitic modes and unstable operation. Mainly two types of mode competition are possible during the gyrotron operation [Edg93]

1. Mode-amplitude interaction: This is an ordinary and most common type of interaction, which is purely amplitude dependent. During gyrotron operation, at any point, the mode having the highest amplitude suppresses the excitation of the other neighboring modes. This also supports the possibility for only single-mode excitation.

2. Phase-amplitude interaction: This type of interaction is associated with the mode competition of a main mode $TE_{m,n}$ with its parasitic satellites, $TE_{m-1,n}$ and $TE_{m+1,n}$. Because of the same radial index, these modes are having same azimuthal bunching. For low-frequency gyrotrons operated with low order modes, the distance of the main mode and the parasitic mode is larger than the gyrotron bandwidth. So, this type of mode competition is not critical for lower order operating modes, but in case of the DEMO gyrotron, the operating mode eigenvalue is more than 100 leading to dense mode spectra. Thus, this

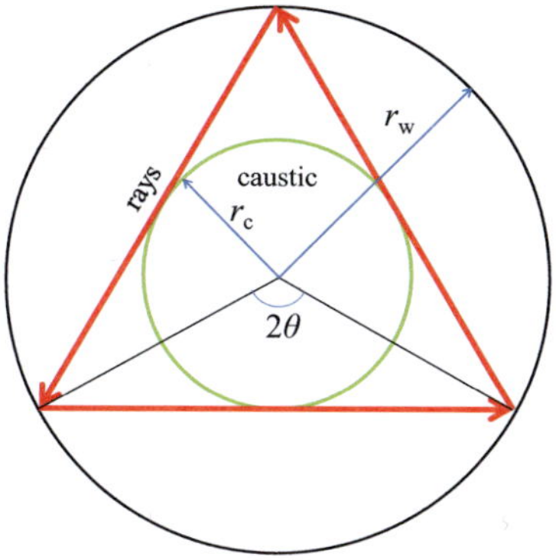

Figure 1.6: Optical description of ray propagation in cylindrical waveguide.

type of interaction is also important for the accurate estimation of the gyrotron behavior. In addition to these parasitic satellites, the modes $TE_{m-3,n+1}$ and $TE_{m-2,n+1}$ are also critical as those lie within a few MHz away from the main mode [Ker96].

1.3.4 Quasi-optical mode converter

The main task of a gyrotron output system is to separate the energetic electron beam from the RF wave and convert the high order complex operating mode into a Gaussian-like mode, which facilitates efficient and low-loss transmission [VZP75]. The output system can be classified into the two parts namely 1. linear output system and 2. transverse output system. In low power application scenario ($P_{out} \leq 50$ kW, CW), the diameter of the collector is smaller than the magnet bore-hole diameter. Simple linear type output system is used, in which the collector serves as an output waveguide. However, in the case of high-power fusion gyrotrons ($P_{out} \geq 50$ kW, CW), transverse-type output systems are used [Mal16].

The quasi-optical launcher works as a waveguide launcher antenna and converts the high order TE mode in to the Gaussian like mode. As next stage, a group of quasi-elliptic and toroidal mirrors are further used for phase correction and beam shaping purposes, which is also essential to control stray radiation inside the tube. In addition to these internal mirrors, an external Matching Optics Unit (MOU) also consist of a set of mirrors which further modifies the RF beam as per the requirements of the transmission line and supply linearly polarized beam.

The initial design of such a launcher has been proposed by Vlasov et al. in [VZP75]. Various advanced concepts were also proposed, i.e. the harmonically deformed launcher [DKM+92], the Kirchhoff–Huygens integral equation based mirror-line launcher [CDK+06] [JTP+09] [JFJ+13], and the hybrid-type launcher, [JTGJ16] [JGJ+15]. The performance of the different design approaches are compared in [JGJ+15].

Inside the overmoded cylindrical waveguide, the high order operating mode can be represented using the method of Geometric Optics (GO) [MT93] [VZP75]. The mode is illustrated as a system of rays, which are successively reflected by the waveguide wall. Relative to the waveguide axis, each mode propagates at the Brillouin angle θ_B, which is defined as

$$\theta_B = \arcsin\left(\frac{\chi_{\mathrm{m,n}}}{kr_{\mathrm{w}}}\right), \tag{1.28}$$

here, r_{w} is the waveguide radius. The minimum distance of the rays from the waveguide center can be calculated using [MT93],

$$r_c = \frac{m}{\chi_{\mathrm{m,n}}} r_{\mathrm{w}}, \tag{1.29}$$

where, the radius r_c is known as the caustic radius. Considering cylindrical symmetry, the distance between two successive reflections can be estimated using

$$L_B = 2r_{\mathrm{w}}\left[1-\left(\frac{m}{\chi_{\mathrm{m,n}}}\right)^2\right]^{1/2}\cot\theta_B, \tag{1.30}$$

and the angle in transverse direction is calculated using,

$$\Delta\phi = 2\theta = 2\arccos\left(\frac{m}{\chi_{\mathrm{m,n}}}\right). \tag{1.31}$$

The inclination of the ray, or in other term, the slot angle of the waveguide can be calculated using [OP09],

$$\psi = \arctan\left(\frac{\theta\tan\theta_{\mathrm{B}}}{\sin\theta}\right). \tag{1.32}$$

The detailed design criteria and performance analysis of a quasi-optical launcher is presented in section 3.3.

1.3.5 Collector

After the beam-wave interaction inside the cavity, the spent electron beam is finally absorbed by the collector. Using the depressed collector concept (single-stage or multi-stage), electrons are decelerated before final absorption and this deceleration energy is recovered by the high voltage power supply. The use of a depressed collector also reduces the cooling requirement and improves the overall lifetime. Considering 1 MW continuous wave operation with typical 35 % interaction efficiency, the energy of the spent electron beam is approximately 2 MW. Thus, the problem of residual energy dissipation is critical for the CW operation and magnetic beam sweeping systems are used to limit the peak power density [IPDR04] [DIP+05]. The Vertical Field Sweeping System (VFSS) and the Transverse Field Sweep System (TFSS) are well known beam sweeping systems

[MSMS04][ESL+07]. The collector is insulated from the main part of the gyrotron. The detailed analysis of depressed collector operation is presented in [SGH+91] [GKVZ97] [PIDT96] [SWRG97] and latest proposal of a multi-stage depressed collectors are presented in [PWI+16] [WPI+16] [PWIJ16] [WPG+17].

The gyrotron interaction efficiency is defined by the equation [Edg93],

$$\eta = \frac{P_{\text{out}}}{P_{\text{beam}}}. \tag{1.33}$$

Here, P_{out} defines the total output power and P_{beam} is the total beam power ($V_b I_b$). Considering the power recovered by collector, $P_c = V_c I$, the collector efficiency is expressed as,

$$\eta_c = \frac{P_c}{P_{\text{beam}} - P_{\text{out}}}, \tag{1.34}$$

and the overall gyrotron efficiency with depressed collector is expressed as,

$$\eta_{\text{total}} = \frac{\eta_0}{1 - (1 - \eta)\eta_c}. \tag{1.35}$$

In the case of DEMO gyrotron, the use of multi-stage depressed collector is mandatory to fulfill the total efficiency requirement of 60 %.

1.4 Major challenges for the DEMO gyrotron design

Based on the design goals and technical constraints presented in the previous section, the major challenges of a high frequency (> 200 GHz), hollow-cavity DEMO gyrotron are discussed here.

Cavity wall-loading

The dependency of the Ohmic loading at the cavity wall (ρ_Ω) (see equation 1.27), on the operating frequency (f) is described as ($\rho_\Omega \propto f^{5/2}$). At the frequency range of DEMO gyrotrons (> 200 GHz), the technical constraint of the cavity wall-loading, limits the selection of the operating modes, beam parameters and eventually, the performance of the tube. The main challenge for a hollow-cavity design is to increase the output power per tube with a desired mode stability.

Mode competition

TE modes having eigenvalues of more than 100 are suggested to meet the power requirement for a DEMO gyrotron, which allows the cavities with large diameter. The mode competition is critical at this eigenvalue range. Due to the dense mode spectra, a DEMO gyrotron demands precise cavity design, optimized operating parameter and suitable start-up condition for the steady-state excitation of the desired mode.

Space-charge neutralization

To support steady state operation of a DEMO plant, the gyrotrons are expected to operate in the continuous wave condition. As compared to the coaxial cavity design, the voltage depression is high in the hollow-cavity gyrotron and the effect of space-charge neutralization is also critical. The space-charge neutralization effect is introduced in section 4.2. Due to this effect, the overall beam energy increases during the CW operation, which may lead to non-stable gyrotron operation. In order to ensure stable CW operation, these effects should be considered in the design process of the DEMO gyrotron to ensure stable CW operation.

Beam misalignment tolerances

In ideal case, the electron beam should be exactly concentric to the gyrotron axis. But due to a misaligned magnetic field and mechanical tolerance in the MIG, an off-axis electron beam may be generated, which affects the beam coupling scenario and mode competition. The performance of a high frequency gyrotron (f > 200 GHz) is greatly influenced by the beam misalignment, because of the very short wavelength.

1.5 Structure of the thesis

The thesis consists of two main parts. Based on the specifications for the advanced DEMO gyrotron and present challenges, a feasibility of the 236 GHz hollow-cavity concept is investigated in the first part (chapters 2, 3 and 4), while in the second part, the operational limits are determined for improved tube performance (chapters 5 and 6).

1.5.1 Part-I: basic feasibility study

In chapter 2, using the suggested systematic cavity design approach, the hollow-cavity design has been finalized and the optimum operating parameters are suggested. The stability of the operation is analyzed rigorously by using multi-mode, time-dependent, self-consistent simulations. Realistic cavity material conductivity is also considered in the analysis. The influence of the beam radius on mode competition has been studied and an optimized beam radius has been selected.

In chapter 3, the possibilities of the fast-frequency, step-tunability and multi-purpose operation are validated with the help of realistic numerical simulations. Efficient MW-level operation within the ± 10 GHz frequency range is demonstrated along with the multi-frequency operation for multi-purpose applications. Using the in-house code TWLDO [JTGJ16] [JGJ+15], a hybrid type quasi-optical launcher has been designed and its performance is validated for the multi-frequency and step-tunable operation.

As the designed hollow-cavity gyrotron operates at a high-order mode (eigenvalue $\sim$ 103) having a dense mode spectrum, the performance of the gyrotron is sensitive to electron beam misalignment. In chapter 4, the beam misalignment tolerance is determined using the realistic mirco-electron based numerical simulations. In the second part of this chapter, the effects of space-charge neutralization has been studied for CW operation. Variations of the beam parameters due to the neutralization are included in the multi-mode, start-up scenario and the stability of the DEMO gyrotron is verified.

1.5.2 Part-II: pushing the limits for the output power

In future fusion plasma experiments, high output power per tube is desirable to reduce the requirement of the total number of tubes per plant. The upper limit of mode eigenvalue can suggest the suitable operating mode with maximum output power and the sufficient mode stability. In chapter 5, the influence of mode competition is studied and the analytical approaches are suggested to determine the mode eigenvalue limit for stable operation. All the discussed approaches suggest the mode eigenvalue limit of 125 for a 236 GHz DEMO gyrotron, assuming diode start-up. Based on this eigenvalue limit, a new 1.5 MW gyrotron design is proposed. In chapter 6, the various approaches to extend the eigenvalue limit are studied and the triode start-up scenario proved as an effective method to control mode competition. Using a triode-type start-up, the possibilities of mode selectivity are also validated.

2 Systematic cavity design approach

In this chapter, the feasibility study for a future 236 GHz hollow-cavity DEMO gyrotron is presented by considering all relevant design targets and technical limitations. Suitable operating modes are selected based on the possibilities of multi-frequency operations. A systematic design approach for the cavity is suggested, followed by rigorous time-dependent simulations, and a stability analysis of the finalized hollow-cavity design.

2.1 Mode selection strategy

The mode selection criteria for high-frequency (> 200 GHz), high-power gyrotrons are based on the requirements for allowable cavity wall-loading, mode competition, multi-frequency operation for multi-purpose applications and frequency step-tunability.

The cavity wall loading scales with $f^{5/2}/[\chi_{m,n}^2 \cdot (1-(r_c/r_{cav})^2)]$ (refer equation 1.26). At high operating frequency (f > 200 GHz), to maintain a wall-loading below an acceptable limit, a possible solution must consider the use of modes with high eigenvalues and/or modes with small caustic radius. These mode properties are satisfied by the asymmetric, high-order volume modes ($m \gg 1$, $n > 2$) with relative caustic radii less than 0.5. As compared to the "whispering gallery" $TE_{m,n}$ modes (with $m \gg n$, n = 1,2), the asymmetric volume modes suffer less from high wall-loading [Bor91]. To realize the power requirement of the DEMO gyrotron with controlled cavity wall-loading (refer Table 1.1), asymmetric modes having eigenvalues greater than 95 are suggested. The upper limit for the mode eigenvalue selection is limited by mode competition. As the mode eigenvalue increases, the mode spectra become more dense. With very high order operating modes, it becomes difficult to excite the desired operating mode due to the mode competition.

The possible approaches to finalize the upper eigenvalue limit for mode selection are discussed with great details in chapter 5. However, for mode eigenvalue greater than 95, these dense mode spectra are nearly identical for nearby modes [FAG+14]. For example in Figure 2.1, the mode spectra of $TE_{-43,15}$ and $TE_{-44,15}$ modes are compared. Within a frequency range of 224 GHz to 260 GHz, all neighboring modes with relative coupling factor higher than 35 % are plotted. Both mode spectra are similar in the considered bandwidth, which makes the selection of a particular high-order mode from a suitable area in the m-p plane less relevant than in the case of lower-order modes [FAG+14]. Because of this, the mode selection is done according to the additional criteria like multi-frequency operation and frequency tunability, which is limited by the RF window design along with the design of the quasi-optical launcher.

2.1.1 Mode selection criteria for multi-frequency operation

As discussed in section 1.2.1, a multi-frequency gyrotron is prerequisite for multi-purpose applications. To support multi-frequency operation, the selected operating frequencies and corresponding cavity modes must fulfill design criteria of the RF window and the quasi-optical launcher. For a single disk RF window, the transparent window frequency can be calculated using,

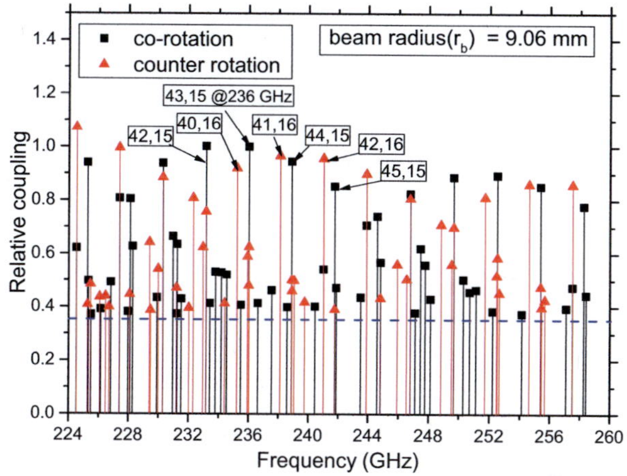

(a)

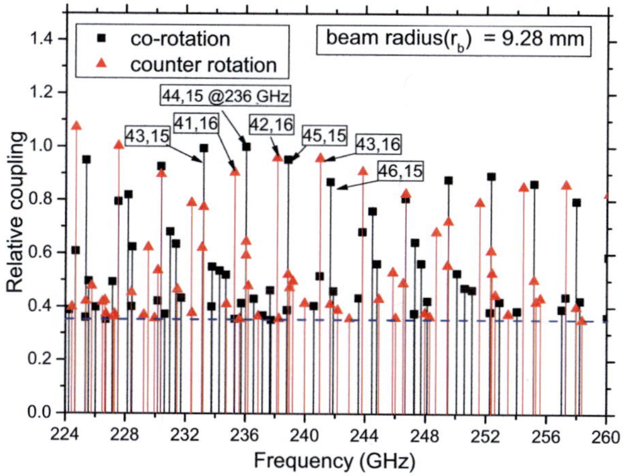

(b)

Figure 2.1: Comparison of the mode spectra in case of the high-order modes. Mode spectra around the cavity operating mode (a) $TE_{-43,15}$ (eigenvalue = 103.21) (b) $TE_{-44,15}$ (eigenvalue = 104.46). The cavity radius is 20.88 mm and 21.14 mm, respectively.

$$f_N = \left(\frac{c}{2 \cdot d \cdot \sqrt{\varepsilon_\mathrm{r}}} \right) \cdot N = f_\mathrm{B} \cdot N, \tag{2.1}$$

where d is the window thickness, ε_r is the dielectric constant of the window and N is an integer. The selected operating frequencies for multi-frequency operation should correspond to the transparent window frequencies for minimum reflection. The minimum and maximum disk thickness can be determined by considering mechanical stability and cost. The typical disk thickness of a CVD-diamond window (ε_r = 5.67) is between 1.50 mm to 2.25 mm, which corresponds to the basic frequencies (f_B) of 42 GHz and 28 GHz, respectively. The RF window with disk thickness of 1.861 mm is selected for this analysis. The reflection and transmission coefficient plots of this window are presented in Figure 2.2. The transmission bands of this window are in the proximity of 104 GHz / 137 GHz / 170 GHz / 203 GHz / 236 GHz / 269 GHz and 304 GHz which support multi-frequency operation at these frequencies.

In addition to the window criteria, the selected modes for multi-frequency operation must have similar caustic radii to support efficient electron beam coupling and mode conversion using the same quasi-optical launcher design [TFK+15] [MT93]. As discussed in section 1.3.4, if the caustic radius is same for the selected modes at various frequencies then, the launcher radius, launcher cut-length, and Brillouin angle also remain unchanged.

Based on the above analysis, the selected operating frequencies and the corresponding modes for the multi frequency operation are listed in Table 2.1 along with their corresponding applications for different aspect ratios of DEMO. The modes for 170 GHz and 203 GHz were suggested and successfully tested by the JAEA gyrotron team [SKO+13] [IOK+16]. This mode series is further extended for the higher frequencies and suitable operating modes are selected for 236 GHz and 269 GHz. The relative caustic radii of all selected modes are nearly the same with only a small deviation of 0.08 % from the average value.

2.2 Systematic cavity design approach

After the suitable mode selection for a 236 GHz DEMO gyrotron, a systematic hollow cavity design approach is suggested as part of this work. The geometry of the hollow-cavity with all the physical design parameters is presented in Figure 2.3. The cavity is a cylindrical-symmetric structure with a straight midsection, a down-taper section and an up-taper section. To the reduce unwanted mode conversion due to abrupt discontinuities, adjacent sections are connected via parabolic smoothing [IWK+95]. As the TE modes operate close to the cut-off frequency, the interaction section radius can be calculated using equation 1.20. For the operating mode $TE_{-43,15}$ at 236 GHz, the calculated cavity radius (r_{cav}) is 20.88 mm. Using equa-

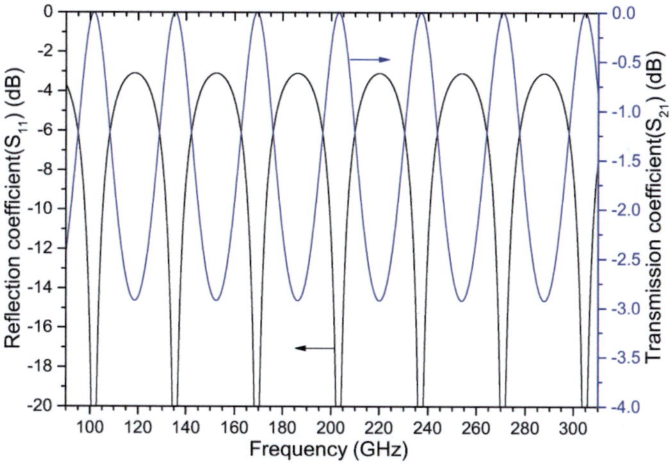

Figure 2.2: Calculated transmission coefficient (S_{21}) and reflection coefficient (S_{11}) of a single disk CVD-diamond window (d = 1.861 mm) within the frequency range of 90 GHz to 310 GHz.

Table 2.1: Applications of the transparent window frequencies and properties of the selected operating modes of a multi-frequency hollow-cavity DEMO gyrotron. (here, H = plasma heating, CD = current drive, A = DEMO aspect ratio)

Window transmission band (-20 dB) (GHz)	168.2 - 170.4	202.0 - 204.2	235.8 - 238.1	269.7 - 271.9
Applications	H (A=3.1)	H (A=3.6) CD (A=3.1)	H (A=4.0) CD (A=3.6)	CD (A=4.0)
Cavity mode	$TE_{-31,11}$	$TE_{-37,13}$	$TE_{-43,15}$	$TE_{-49,17}$
Mode eigenvalue	74.32	88.76	103.21	117.65
Relative caustic radius	0.4171	0.4168	0.4166	0.4165
Normalized window thickness	5/2	6/2	7/2	8/2

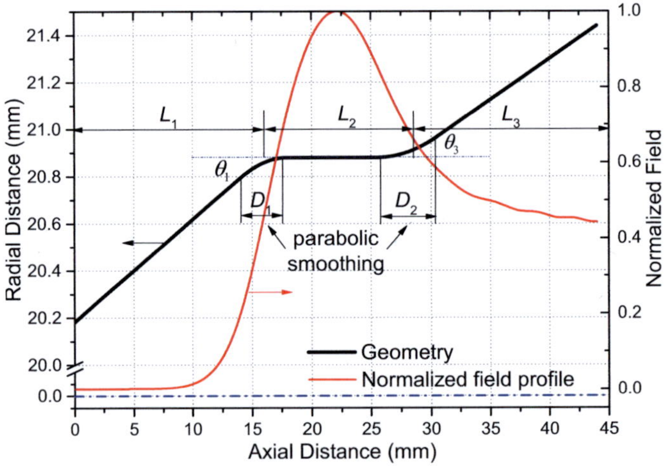

Figure 2.3: Geometry of the hollow cavity design and the normalized magnitude of the longitudinal field profile of the operating mode.

tion 1.22, the electron beam radius of 9.06 mm is calculated for maximum coupling. As a next step, based on the desired quality factor and axial field profile, all other physical parameters of cavity have to be optimized.

After finalizing the cavity radius, the parametric analysis for the interaction length (L_2) is performed. At the beginning, all other geometry parameters are chosen from previous experiences for the cavity design of the W7-X (140 GHz) and ITER (170 GHz) gyrotrons. The initially selected values of these parameters were: $L_1 = L_3 = 16$ mm, $\theta_1 = \theta_3 = 2.5\,^\circ$, $D_1 = D_2 = 5$ mm. The parametric analysis of the interaction length L_2 is presented in Figure 2.4. For different interaction section length, the gyrotron performances are determined with the help of the Self-Consistent (Self-C) code of the in-house code package termed "Cavity" [Ker96]. From present consideration of cavity cooling technology, the wall-loading limitation of 2 kW/cm^2 is selected for this analysis. Based on the maximum cavity wall-loading of 2 kW/cm^2, the operating parameters for each interaction section length are finalized and the results of these activities are summarized in Table 2.2.

Along with the diffractive quality factor, the Ohmic wall-loading also increase with length. Thus, to keep constant wall-loading, it is necessary to reduce the electron beam power, which consequently reduces the output power. The self-consistent time-dependent analysis with the consideration of main competing modes recommends a mid-section length of less than 13 mm. For higher mid-section lengths, the effect of mode competition become critical and it is difficult to excite the desired operating mode. After this analysis, the interaction length (L_2) of 12 mm is selected for further analysis, leading to a good compromise between high output power, high efficiency and acceptable mode competition.

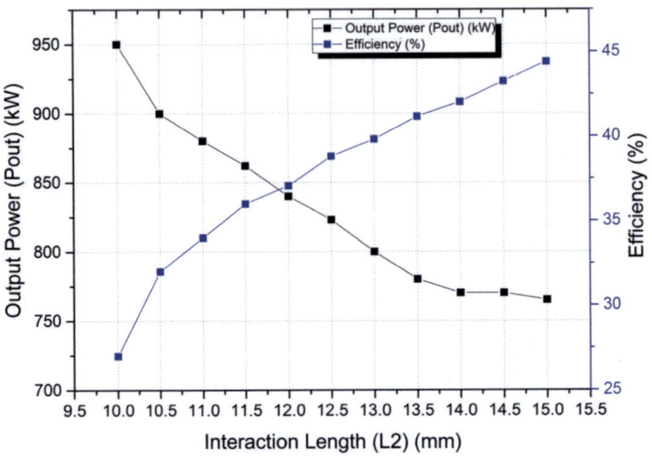

Figure 2.4: Output power and efficiency of the hollow-cavity design with various interaction section lengths L_2 from 10 mm to 14 mm. The operating parameters are selected for maximum efficiency with wall-loading limitation of 2 kW/cm^2.

As a next step, the effect of the input and output angles on the gyrotron performance is analyzed. For L_2 = 12 mm and corresponding operating parameters from Table 2.2, output power, efficiency and wall-loading are plotted in Figure 2.5, as a function of input and output angles.

As compared to the initial value of the output angle of 2.5 °, the output power can be increased from 811 kW to 824 kW by using an output angle of 2 °. The optimum point in terms of the maximum power and efficiency with wall-loading less than 2 kW/cm^2, is illustrated in Figure 2.5.

As a last step of the cavity design, the input and output taper lengths (L_1 and L_3) and smoothing regions (D_1 and D_2) are optimized using the scattering matrix codes. The input-taper section is designed in such a way that the operating frequency is lower than the cut-off frequency.

Table 2.2: Performance and operating parameters of the cavity for several midsection lengths (L_2).

Interaction length (L_2) (mm)	10	10.5	11	11.5	12	12.5	13	13	14	14.5	15
Diffractive quality factor (Q_{diff})	1007	1131	1220	1363	1465	1628	1743	1927	2057	2263	2409
Magnetic field (B_0) (T)	9.36	9.20	9.16	9.14	9.12	9.10	9.08	9.07	9.05	9.04	9.02
Beam current (I_b)(A)	47	44	42	40	39	36.5	35.5	34	33.5	33	32.5
Beam energy (keV)	72	62	60	59	58	56.5	55.5	54.5	53.5	53	52
Ohmic wall-loading (kW/cm^2)	2.01	2	2	1.99	1.99	2	1.98	1.98	1.98	2	2
Output power (kW)	950	900	880	862	840	823	800	780	770	770	765

Thus, it works as a reflector and helps the cavity to maintain the desired Q value. Using the scattering matrix code of the "Cavity" code-package [Ker96], the mode transmission properties in the input-taper and output-taper section are determined. With the original design (L_1 = 16 mm, D_1 = 4 mm), more than 99 % of the $TE_{-43,15}$ mode amplitude is reflected from the input section without any spurious mode excitation. Concerning the up-taper section, more than 99 % of the $TE_{-43,15}$ is transmitted using L_3 = 16 mm, D_2 = 5 mm.

The step-wise description of the proposed systematic cavity design approach is summarized in Table 2.3. Using the systematic cavity design approach, the finalized cavity parameters are shown in Table 2.4. The final geometry with the longitudinal field profile is presented in Figure 2.3. The diffractive quality factor (Q_{diff}) and the Ohmic quality factor (Q_Ω) of the final cavity design is 1443 and 62937, respectively. In order to study the influence of individual cavity parameter on the output power and efficiency, the performance of the cavity is verified by variation of the physical parameters. At a time, only one physical parameter is varied and the final results are presented in Table 2.5. In this case, equation 2.2 provides the definition of the percentage variation p of quantity X (output power P_{out} or efficiency η).

$$p_X = \frac{2 \cdot (X_{max} - X_{min})}{X_{max} + X_{min}} \cdot 100\ \% \tag{2.2}$$

However, the total influence on the cavity performance is not simply equal to the sum of the individual ones, when more than one physical parameters are varied simultaneously. As described in Table 2.5, the gyrotron performance is strongly influenced by the midsection length L_2, making it the first parameter to be addressed while designing the cavity.

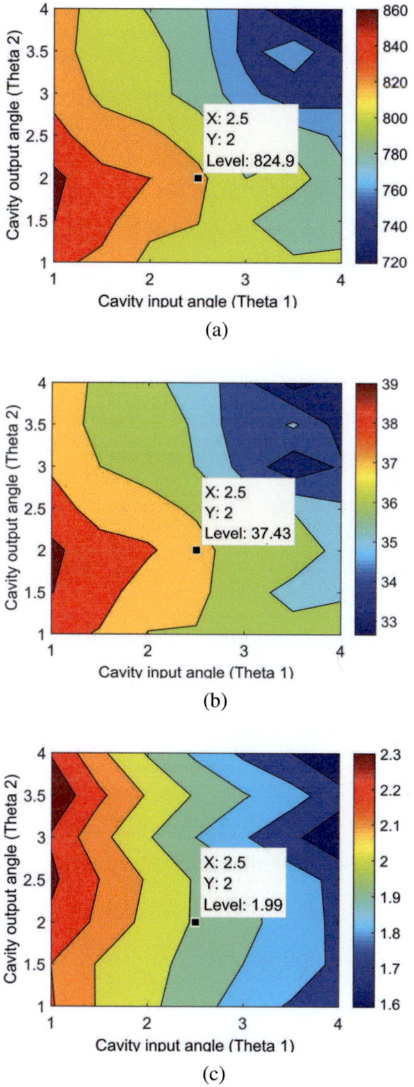

Figure 2.5: (a) Output power, (b) efficiency and (c) Ohmic wall loading at various input and output angles of cavity. For cavity wall-loading of 2 kW/cm^2, the optimum point is indicated in each diagram with its corresponding value.

Table 2.3: Steps of the systematic cavity design approach [KAF+16c].

Step1: Calculate mid-section radius and beam radius using standard formulas for the desired operating mode.
Step 2: Estimate initial geometrical cavity parameters (L_1, L_2, L_3, θ_1, θ_2, D_1, D_2) from the design goals, technical limitations, cold cavity profile and previous cavity design experiences.
Step 3: Simulate gyrotron interaction with various mid-section lengths (L_2), keeping all other physical parameters fixed, and find the optimum value (e.g. Figure 2.4 and Table 2.2). The operating parameters (beam voltage, beam current, magnetic field, etc.) must be optimized for each particular L_2.
Step 4: Optimize input and output angles (θ_1, θ_2) with regards to power and efficiency using the optimized mid-section length from step 3 and corresponding operating parameters (e.g. Figure 2.5).
Step 5: Optimize input taper section length (L_1) and input smoothing length (D_1) by calculating reflections of the RF modes from input section. These lengths should be set to obtain maximum reflection of the operating mode without mode conversion.
Step 6: Optimize output taper section length (L_3) and output smoothing length (D_2) to have maximum RF power transmission to the quasi-optical launcher without mode conversion.

Table 2.4: Final physical parameters of the 236 GHz hollow cavity design.

Input taper	
Length L_1	16 mm
Angle θ_1	2.5 °
Smoothing length D_1	4 mm
Midsection	
Length L_2	12 mm
Radius r_{cav}	20.88 mm
Output taper	
Length L_3	16 mm
Angle θ_2	2°
Smoothing length D_2	5 mm

Table 2.5: Effect of the individual physical parameters on the overall cavity performance.

Parameter	L_2	θ_1 and θ_2	D_1 and D_2	L_1 and L_3
Investigated range	10 – 15 mm	1 ° – 4 °	1 – 4 mm	10 – 19 mm
Output power variation p_X	21.57 %	17.72 %	14.01 %	3.68 %
Efficiency variation p_X	44.0 %	16.7 %	12 %	4 %

2.3 Multi-mode time-dependent simulations

2.3.1 Set-up for time-dependent cavity simulations

The various approaches for the slow-time scale, Self-Consistent (Self-C) calculations are presented in [Edg93] (chapter 3, 4 and reference within). In this work, the in-house time-dependent, multi-mode, self-consistent code-package EURIDICE [APIV12] [Avr15] is extensively used for the interaction simulations. The features of the EURIDICE code package are discussed in section A.2. Over the time, the results are further verified with the in-house code SELFT from the code-package "Cavity" [Ker96]. The behavior of the mentioned codes are also successfully benchmarked with the experimental results along with the results from other numerical codes. For example, in Appendix A.5, for the test-case of W7-X and ITER gyrotrons, the performance of SELFT and EURIDICE codes is compared with the TWANG and TWANG-PIC codes, which are developed at the Swiss Plasma Center - École Polytechnique Fédérale de Lausanne (SPC-EPFL) [ATA$^+$11] [BTV$^+$15][BTV$^+$14]. The performance comparison of the SELFT code and EURIDICE code is also presented in [Sch15].

In addition to these codes, the interaction code developed at the Technical University Hamburg-Harburg (TUHH) [JGS99] [Jel00] , University of Maryland (MAGY- MAryland GYrotron) [BAL$^+$98] [VAC$^+$08] and Indian Institute of Technology Roorkee (GDS- Gyrotron Development Suite) [JK10] are highly recognized for the gyrotron cavity design and analysis. In various studies, the commercial Particle-in-cell (PIC) based codes, 2D/3D MAGIC [MSBL98], [BKC99], [KSK$^+$11], [LHC$^+$10], CST [CJ12], [CST] and VSIM [LS13][LSC$^+$14] are used. However, in the case of high-power gyrotrons with very high-order operating modes (eigenvalue $\sim$ 100), the use of such codes is under discussion, due to the need of excessive computational resources. In [SKL$^+$14] [SKH$^+$14], the performance comparison of the PIC-based commercial codes is presented with experimental verification. The setup for the multi-mode, self-consistent simulation using EURIDICE code package is discussed here. As a pre-process, the geometry file is created using the desired physical cavity parameters, which is further used for the interaction simulation. The resonator axis is discretized in very small steps (normally in the steps of 100 μm to 200 μm) and considering the azimuthal symmetry, each axial point is link to the respective value of resonator radius. In the case of coaxial-type cavity, insert information shall be also included in the geometry file.

Based on the selected frequency range and cavity profile, the list of TE modes are selected using Coldcav code from the EURIDICE code package, which are the possible neighboring modes for the analysis. The mode eigenvalue can be calculated using the "CHIMP" module as discussed in [Ker96]. The frequency, mode coupling, quality factor and axial field-profile for each individual modes are calculated without considering the effects of electron beam. The wave equation is solved with the appropriate boundary conditions. For the 236 GHz DEMO gyrotron, the simulated axial field-profile of the main mode $TE_{-43,15}$ is plotted in Figure 2.3, while the neighboring modes and their relative mode coupling is shown in Figure 2.1 (a).

Considering time-varying beam energy and/or magnetic field, the multi-mode time-dependent, self-consistent simulation can be performed using Evridiki code. The equations of electron motion (2nd order Predictor-Corrector scheme) and electromagnetic field (Crank-Nicolson finite difference approach) are solved self-consistently. Both trajectory and quasi-PIC approaches are implemented. Using multi-mode, start-up scenario simulations, the possible mode excitation can be estimated at specific time. The possible mode competition in gyrotron interaction is discussed in section 1.3.3. To include the effects of possible neighboring modes, up to 99 parasitic modes can be incorporated in the multi-mode simulations. To reduce computational time, the computation is parallelized using the Message Passing Interface (MPI). During the start-up simulations, variation of beam energy can be selected either linearly or step-wise. As per the variation of beam energy, the beam current and velocity ratio either changes according to the adiabatic approximation and the Schottky formula or defined by the user externally. Unlike fixed-field approximation, the axial variation of field profile is also considered in the simulation.

2.3.2 Performance of the suggested cavity design

In this section, an RF behavior of the proposed cavity design is investigated to further validate the selected operating parameters and design. As the 236 GHz gyrotron operates in very high-order modes with dense mode spectra, the selection of the operating parameters and start-up conditions is critical for the stable operation. Using single-mode simulations, the gyrotron operating parameters are finalized with the criteria of more than 35 % interaction efficiency, high output power and stable RF operation. Since the magnetic system for the DEMO gyrotron is not finalized yet, therefore, a constant magnetic field profile is used for the analysis. At this stage of the analysis, the typical cavity conductivity (σ) of 1.4×10^7 S/m is selected. The optimized parameters from the single mode simulations are: beam energy = 58 keV, beam current = 39 A, magnetic field = 9.130 T.

To verify the effect of competing modes on mode stability, multi-mode time-dependent simulations were performed as next step. Within a frequency range of -5 % to +10 %, all the modes having relative coupling coefficient higher than 0.35 are included in the multi-mode analysis. The mode spectrum of the selected neighboring modes is presented in Figure 2.1 (a). For the proposed cavity design, the result of the multi-mode, time-dependent simulation with the 99 neighboring modes is presented in Figure 2.6. Along with time, the beam energy changes linearly from 20 keV (noise level) to

58 keV, while velocity ratio (α) and beam current (I_{b}) are varying adiabatically to reach their final value of α = 1.25 and I_{b} = 43 A, respectively. As the implementation is based on the self-consistent approach, the change in the axial-profiles for each mode is also incorporated.

For reliable simulations, the convergence analysis has been performed for two important numerical parameters: (a) number of particles for macro-electron beam and (b) time step for the simulation. The number of electrons for macro-electron beam should be sufficient to characterized interaction between electron beam and field. In this work, a minimum of 19 beamlets have been selected with a total number of particles more than 400. At each time-step, the random initial values of electrons are assigned to control non-physical behaviors [Nus81]. The transit time of the electrons along the cavity (< 1 ps) is much lower than the evolution time of the field [APIV12], which supports the constant field profile during the transition time of electrons. Considering the field variation time of the order of 0.1 -

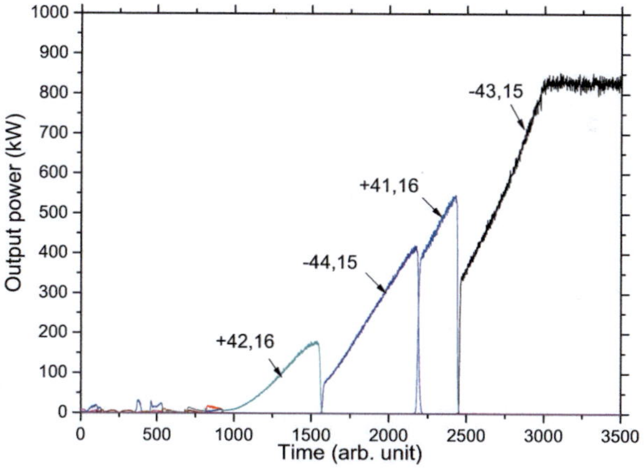

Figure 2.6: Multi-mode, time-dependent simulation with ideal beam parameters. (V_b = 20 - 58 keV, I_b = 39 A, B_0 = 9.130 T). 99 neighboring modes have been considered for this simulation.

0.2 ns, the time step of 0.01 ns is selected for reliable simulation [Ber10]. The total simulation time is fixed by considering beam energy variation of 10 - 20 V/ns. This time step is two orders above the experimental value [Sch15]. So, in Figure 2.6, the time scale during the start-up is considered in arbitrary unit. The beam energy is increased linearly from 20 keV (t = 0) to 58 keV (t = 3000) and steady-state condition is selected from the time-interval t = 3000 to t = 3500. Despite of the consideration of the large number of neighboring modes, the stable operation of 828 kW is achieved with the interaction efficiency of 38 % using the suggested operating parameters.

2.3.3 Inclusion of realistic electron beam parameters

The gyrotron interaction is highly dependent on the quality of the electron beam. Due to practical limitations of the MIG design, it is not possible to generate an energetic electron beam with ideal beam parameters [Tsi01]. For the accurate performance estimation, the influence of realistic beam parameters on mode stability and interaction efficiency has been investigated. For a 236 GHz DEMO gyrotron, the effects of beam velocity spread and radial width on the beam-wave interaction are studied with the help of multi-mode simulations.

Velocity spread

The total electron energy is distributed differently between the axial and transverse component of the motion. The spread in perpendicular electron velocities can introduce non-uniform spread in the cyclotron frequency, which change the resonance conditions [Nus04] [But84] and, eventually, the interaction efficiency decreases. Even for the modes operating near cut-off, the spread in axial velocity causes a difference in electron transit time along the cavity and reduces the overall efficiency. For efficient gyrotron operation, it is desirable to have maximum orbital-to-axial velocity ratio (α), but the velocity spread of the beam also limits that. In the case of an electron beam, which is having significant spread, as the magnetic field increases along the gyrotron axis, the electrons with small axial-velocity further decrease the axial velocity and are trapped between the gun and resonator region. This phenomena is also known as "magnetic mirror effect" [Edg93].

The possible sources for the velocity spread in the gyrotron electron beam and their analytic formulations are discussed in [Tsi01] [KLTZ92]. However, for gyrotrons operating in the sub-millimeter wave/ millimeter frequency range, the surface roughness of emitter is the main source of the beam velocity spread [KM16]. The effects of the emitter surface roughness on the high frequency fusion gyrotrons have been systematically studied in [ZIP+15]. Various experiments [PIT16] [GGK+99] [Xiz84] and numerical studies [DK99] [ZZM06] [XCYH14] suggest, Gaussian like velocity distribution of the gyrotron electron beam. The same is also considered for this analysis. From the perpendicular velocity spread, the equivalent spread in velocity ratio (α) can be estimated using equation 1.12.

Table 2.6: Influence of velocity spread on gyrotron performance.

$\delta\beta_\perp$ (rms) (%)	$\delta\alpha$ (rms) (%)	Output power (P_{out}) (kW)	Efficiency (η_{int})(%)
0	0	828	38.0
6	15.4	780	36.0
8	20.5	744	34.7
10	25.63	700	33.4

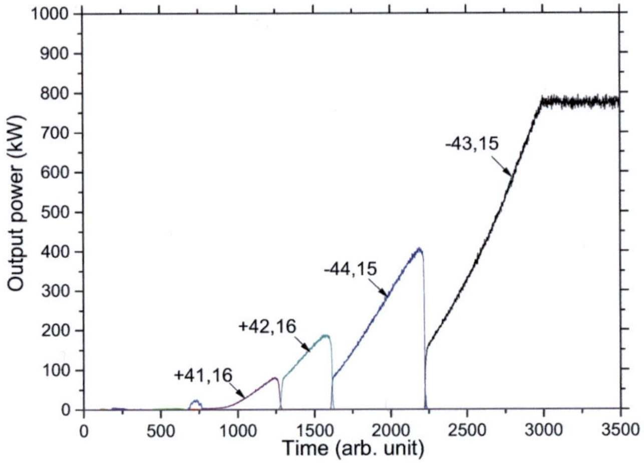

Figure 2.7: Start-up scenario considering 99 neighboring modes and a realistic electron beam with 6 % (rms) velocity spread.

Multi-mode time-dependent simulations have been performed with various perpendicular velocity spreads. Using the in-built function of the EURIDICE code, electron beams having particular perpendicular velocity spread have been generated and selected for start-up simulations. The performance of a 236 GHz DEMO gyrotron with various beam velocity spread is summarized in Table 2.6. The output power and efficiency decreases with increasing velocity spread, but the excitation of the mode remains stable. The recent gun design studies presented in [ZIPJ13] [FAP+15] suggest approximately 6 % of the perpendicular velocity spread for a DEMO gyrotrons. The result of the multi-mode, start-up scenario with 6 % velocity spread is presented in the Figure 2.7. Except the consideration of velocity spread, the simulation setup is same as discussed in section 2.3.2. With the inclusion of 6 % (rms) velocity spread, output power and interaction efficiency are reduced from 828 kW, 38 % to 780 kW, 36 %. Compared to the start-up scenario with an ideal electron beam (Figure 2.6), the excitation of next parasitic mode $TE_{+41,16}$ is suppressed due to the influence of perpendicular velocity spread. The mode competition control by the velocity spread is also supported in [AEZS77] [ZZM06].

Radial width

Gyrotron electron beams are generated by MIGs having a finite width of the emitter. The MIG operates in a temperature limited region and each electron experiences a different static field, which give rise to the generation of the electron beam with radial width. High-frequency (>100 GHz) fusion gyrotrons operate with high order TE modes with dense mode spectra. Because of the very short wavelength and large number of neighboring modes, the radial width of the electron beam plays a significant role in mode excitation and stability.

The performance of the designed cavity is verified for different beam widths. The value of a radial width represents the single side guiding center spread. According to the desired radial width, the maximum and minimum guiding center radii are calculated. Within those radial limits, the electron guiding centers are distributed randomly. The radial width implementation approach

Table 2.7: Effect of a radial width on gyrotron behavior.

Radial width	Radial width	Output power (P_{out}) (kW)	Efficiency (η_0)(%)
$2 \cdot r_L$	$\lambda/8.88$	827	37.72
$3 \cdot r_L$	$\lambda/5.92$	821	37.48
$4 \cdot r_L$	$\lambda/4.44$	815	37.15
$5 \cdot r_L$	$\lambda/3.55$	801	36.46
$6 \cdot r_L$	$\lambda/2.96$	unstable mode	

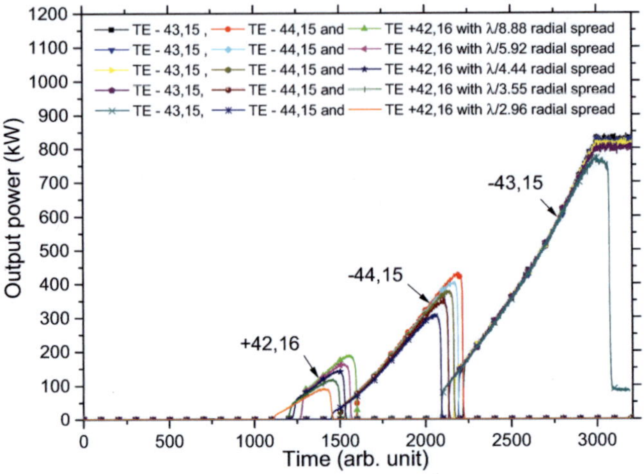

Figure 2.8: Influence of beam radial width on RF behavior of a 236 GHz DEMO gyrotron.

is discussed in section A.3.3. The output power and efficiency with different radial widths of the electron beam are shown in Table 2.7 and the start-up scenarios are shown in Figure 2.8. In these simulations, a concentric electron beam with given radial width is considered, without velocity spread and beam misalignment. The value for the Larmor radius (r_L) is 71.4 μm and the wavelength (λ) is 1.271 mm. It is clear from the results that, the output power and efficiency are not affected significantly with consideration of radial width. However, the operating mode become unstable with the radial beam width of $\lambda/2.96$.

2.3.4 Stability analysis

To ensure reliable operation during the actual experiments, it is requested to have stable operation not only with the optimized parameters, but also within a certain range of the operational parameters. In this part, stability of the operation is verified with respect to electron beam energy. As a first step, the stability margin with the selected operating parameters has been investigated, and in a second step, new operating parameters are suggested to achieve the desired stability margin of 2 keV.

For the selected cavity design and operating parameters, the possible factors of the mode loss at high beam energy has been studied in [KAF+16b]. The results verified that the mode loss is only due to the detuning effect, not because of mode competition. The effects of detuning on the mode stability is now further verified. As shown in Figure 2.9 (a), using a multi-mode start-up scenario, the stable operation is obtained till t = 5000 with the nominal beam parameters (B_0 = 9.130 T, V_b = 58 keV, I_b = 39 A). As next stage, the energy has been increased in steps of 0.1 keV from 58 keV until mode loss. With the selected magnetic field of 9.130 T, the operating mode is stable up to the electron beam energy of 59.0 keV, which is a margin of 1 keV with respect to the nominal beam energy of 58.0 keV.

The stability margin of mode excitation is further extended by operating the gyrotron at higher magnetic field, which corresponds to a lower detuning. Considering a modified magnetic field of 9.145 T at the cavity center, the mode excitation is stable up to 60.4 keV and the desired stability margin of up to 2 keV can be achieved. (see Figure 2.9 (b)). This, of course, is accomplished by a loss of power and efficiency at the operating point, which are now reduced by 760 kW and 35 %, respectively.

2.4 Additional considerations for RF behavior analysis

2.4.1 Selection of optimized beam radius

In this section, the influence of an electron beam radius on the gyrotron operation has been studied. In the case of a 236 GHz DEMO gyrotron, the calculated beam radius for the maximum coupling is 9.06 mm. In [CMM+06],

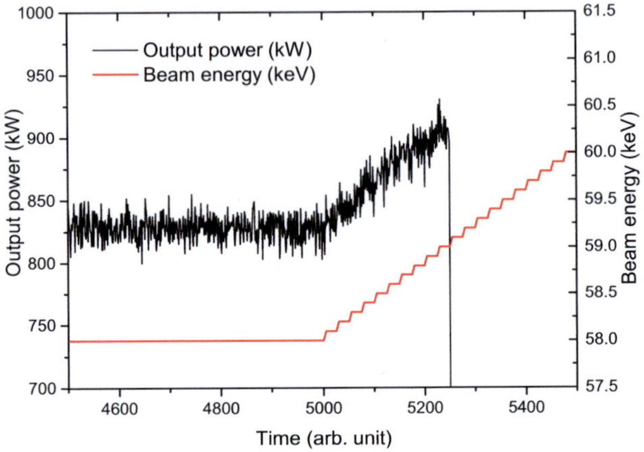

(a)

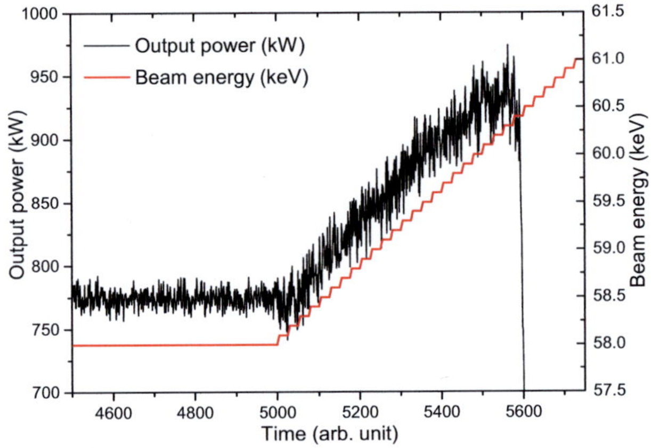

(b)

Figure 2.9: Stability analysis of the selected operating point with (a) magnetic field of 9.13 T and (b) with magnetic field of 9.145 T. The beam energy has been increased in 0.1 keV steps. Using the magnetic field of 9.13 T, the output is stable till 59 keV with the stability margin of 1 keV, The desired stability margin of 2 keV is achieved with a magnetic field of 9.145 T.

with the help of fixed-field calculations, the effects of beam radius on gyrotron mode completion has been studied and it has been proved that the mode competition can be controlled using an electron beam with large radius. In this section, using multi-mode, time dependent, self-consistent simulations, effects of beam radius on mode competition and start-up scenario are studied and optimum beam radius has been selected for a 236 GHz, $TE_{-43,15}$ mode DEMO gyrotron.

Mode competition control

In Figure 2.10, the mode spectra of a 236 GHz DEMO gyrotron are presented with beam radii of 9.06 mm and 9.27 mm. The relative mode coupling can be calculated using equation 1.21 and multiplied with the respective quality factor. The neighboring modes within -5 % to +10 % of frequency range and having a relative coupling of more than 0.5 are considered. With increase in beam radius, the relative coupling of the next counter-rotating modes ($TE_{+m-3,n+1}$ and $TE_{+m-2,n+1}$) decreases, which is a favorable condition for the stable excitation of the operating mode. It has also been noted that, as the beam radius increases, the coupling of counter-rotating $TE_{+43,15}$ also increases. With the beam radius of 9.27 mm, the coupling of the co-rotating mode is equal to the coupling of counter-rotating mode. This condition sets an upper limit for the choice of the possible beam radius. The results of multi-mode, time-dependent, self-consistent simulations with beam radii of 9.06 mm, 9.10 mm and 9.16 mm are compared in Figure 2.11 and 2.12. The beam energy is linearly varied from 35 keV to 62 keV and mode loss is analyzed in each case.

Considering a stability margin of 2 keV, the updated magnetic field of 9.145 T is selected for the simulation. Using the beam radius of 9.10 mm, coupling with the counter-rotating mode $TE_{+41,16}$ is reduced and its excita-

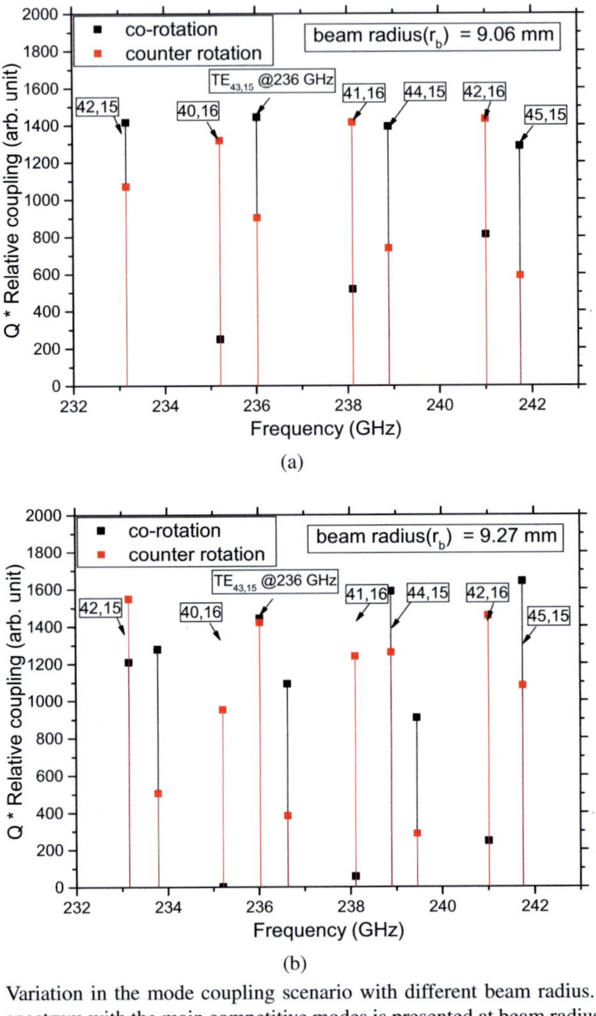

Figure 2.10: Variation in the mode coupling scenario with different beam radius. The mode spectrum with the main competitive modes is presented at beam radius of (a) 9.06 mm and (b) 9.27 mm.

tion can be suppressed during start-up. However, the stability margin of the main mode ($TE_{-43,15}$) is not changed with the beam radius of 9.10 mm. By comparing Figure 2.12 (a) and Figure 2.12 (b), it is clear that the stability of the operating mode decreases from 60.7 keV to 60.5 keV with the beam radius of 9.16 mm.

Using single-mode simulations, the output power of the gyrotron with various beam radius is presented in Figure 2.13. As the beam radius deviates from its optimum value, the output power decreases due to the reduction in beam coupling with the main mode. After considering start-up scenarios, mode stability and coupling, the beam radius of 9.10 mm are suggested, which improves the mode competition scenario with reduced coupling of the counter-rotating mode and marginally reduced output power.

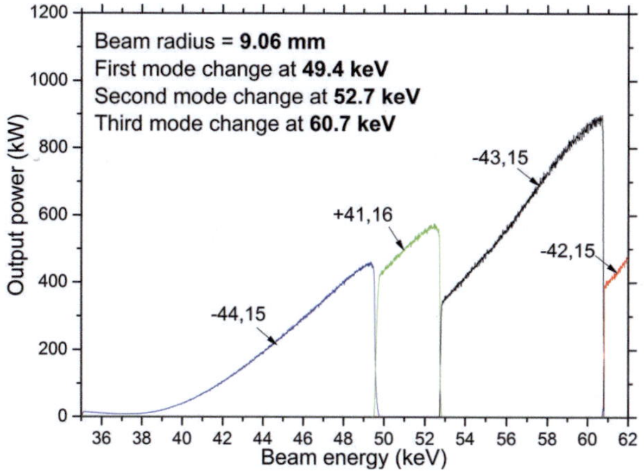

Figure 2.11: Multi-mode start-up scenario with the beam radius of 9.06 mm.

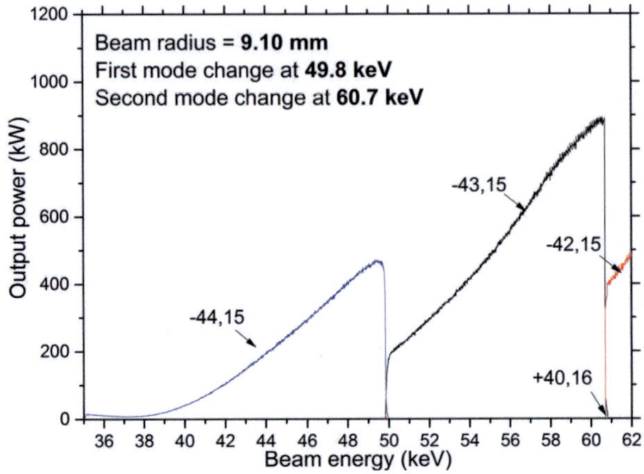

(a)

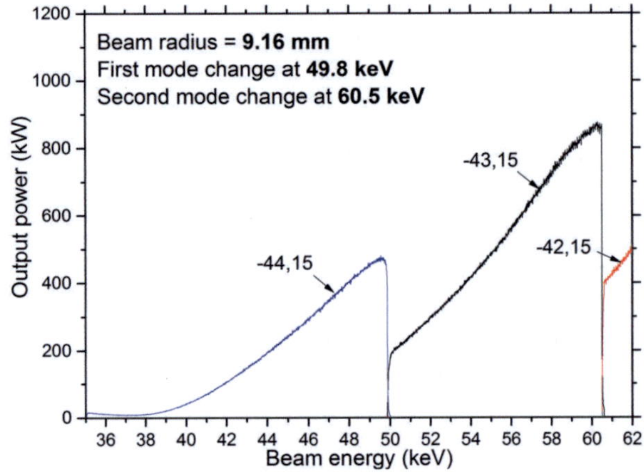

(b)

Figure 2.12: Multi-mode start-up scenario with the beam radii of (a) 9.10 mm and (b) 9.16 mm.

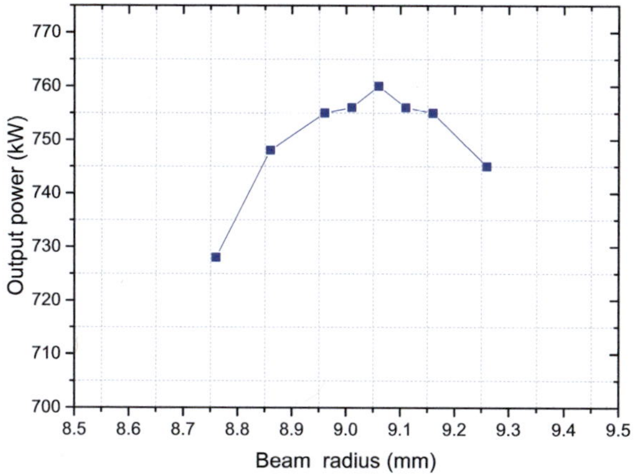

Figure 2.13: Influence of a beam radius on a 236 GHz gyrotron operation.

2.4.2 Cavity performance using realistic material properties

In most of the cases, a gyrotron cavity is manufactured using copper, because of the material's high electrical and thermal conductivity. Glidcop refers to the family of copper based matrix composite alloys mixed primarily with aluminium oxide ceramic particles. It has higher resistance to thermal softening and radiation damage than oxygen free copper with a slight reduction in the effective conductivity. Due to these unique advantages, as of now, Glidcop is the best choice for gyrotron cavity. Depending on the content of aluminium oxide, Glidcop is classified into the three main types: AL-15, AL-25 and AL-60. Increasing the level of aluminium oxide will increase mechanical strength but decrease conductivity. The properties of Glidcop are discussed in [Sho93]. The successful operation of Glidcop cavities is experimentally demonstrated in [RRP$^+$16], [ILN$^+$01], [KIS$^+$12], [CBF$^+$07].

For a standard gyrotron design approach, the typical value of 1.4×10^7 S/m is used as the conductivity of the cavity. Using the detailed formulation discussed in section A.1, it is possible to calculate, the frequency, temperature and surface roughness dependent Glidcop conductivity. With operating temperature of 250 °C and surface roughness of 0.1 µm, the effective Glidcop conductivity for the 236 GHz gyrotron cavity is 1.9×10^7 S/m.

Due to the consideration of high effective conductivity, it is possible to increase the cavity power for the same wall-loading of 2 kW/cm^2 (equation 1.27). The updated operating parameters and cavity performance with the modified conductivity is presented in Table 2.8. The appropriate operating parameters are selected to get at least 36 % of interaction efficiency with the realistic beam properties (6 % velocity spread, $\lambda/4$ radial width). All the physical parameters of the cavity are the same as for the previously

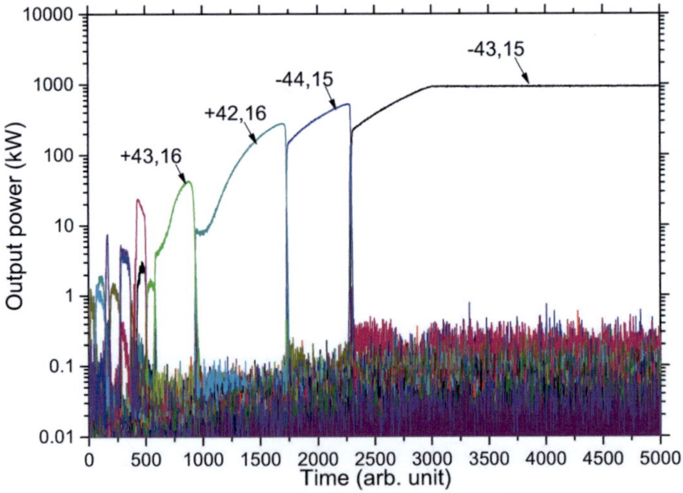

Figure 2.14: Start-up scenario for longer time duration, assuming a realistic Glidcop conductivity of 1.9×10^7 S/m.

Table 2.8: Operating parameters and performance analysis of the DEMO gyrotron using realistic Glidcop conductivity of 1.9×10^7 S/m.

Frequency (GHz)	236.1
Mode	$TE_{-43,15}$
Mode eigenvalue	103.213
Magnetic field (T)	9.165
Beam radius (mm)	9.06
Beam electron energy (keV)	61
Beam current (A)	43
Diffractive quality factor (Q_{diff})	1443
Ohmic wall loading (kW/cm^2)	2.00
Effective conductivity (S/m)	1.9×10^7
Without velocity spread or radial spread consideration	
Output power (kW)	960
Interaction efficiency (%)	38
With 6 % velocity spread and $\lambda/4$ radial width	
Output power (kW)	920
Interaction efficiency (%)	36

Table 2.9: Operating parameters and gyrotron performance with high wall-loading (without realistic beam parameters).

Wall-loading (kW/cm^2)	Beam energy (keV)	Beam current (A)	Output power (kW)	Efficiency (%)
2	61	43	960	38
2.18	65	45	1070	38
2.43	65	50	1200	38

described cavity in Table 2.4. Compared to the standard design with the typical value of conductivity (1.4×10^7 S/m), the output power increases from 778 kW to 920 kW with modified conductivity.

The multi-mode time-dependent simulations with the updated beam parameters are presented in Figure 2.14. Beam energy linearly increases from 20 keV to 61 keV till t = 3000 and remains constant till t = 5000. During the steady-state operation, the powers in the selected neighboring modes are less than the 0.1 % of the main mode power. This result indicates the stable operation without spurious mode generation or mode loss. As discussed in section 2.3.4, the stability analysis with the updated beam parameters has been performed at a magnetic field of 9.175 T are suggests to have stability margin of 2 keV. The effect of cavity wall-loading on the performance of the hollow-cavity gyrotron is illustrated in Table 2.9.

For the same cavity design, the operating parameters are finalized for the high cavity wall-loading. The ideal beam parameters have been considered with the interaction efficiency of 38 %. The output power is approximately proportional to the maximum cavity wall-loading. Here, effective cooling systems support operation with high wall-loading, which increases output power per tube. Thus, an effective cavity cooling is essential for high-power gyrotron operation. The operation with more than 1 MW output power is possible assuming wall-loading of 2.18 kW/cm^2.

3 Fast frequency step-tunability and multi-purpose operation

In the previous chapter, the gyrotron cavity and operating parameters have been finalized considering the stable and efficient operation at 236 GHz. As discussed in section 1.2.1, the DEMO gyrotron requirements also include fast frequency step-tunability and multi-frequency operation for plasma stability control and a multi-purpose gyrotron application, respectively. In this chapter, the possibilities for fast-frequency tunability and multi-purpose operation for the hollow-cavity DEMO gyrotron are investigated.

3.1 Fast frequency step-tunability

3.1.1 Importance of fast frequency step-tunability in fusion plasma experiments

In ideal conditions, the tokamak plasma is toroidally symmetric and all field lines on a particular magnetic flux surface carry the same current. With the increase in plasma density, the strong affect in the plasma current profile is observed, due to the atomic processes and radiation cooling at the edge. In this case, some magnetic field lines carry more current than others. This increased plasma current, also known as 'Bootstrap current', flows along the magnetic lines, which causes corrugations in the magnetic flux surface and generates magnetic islands. These magnetic islands reduce the overall plasma pressure and eventually, decrease the fusion energy and plant efficiency.

The bootstrap current is proportional to the pressure gradient and, it should be removed from magnetic islands. This reduction in the bootstrap current further increases the size of the magnetic islands. The resulting large magnetic islands increase the toroidal transport, which is also called as Neoclassical Tearing Mode (NTM), and leads to major loss of confinement and disruption [Igo15] [BGG+00]. This type of instability can be avoided by driving current in the magnetic island. Electron Cyclotron Current Drive (ECCD) is the preferred method for NTM stabilization due to its precise localization of the power deposition [Mor92] [FRC+12].

The position of such instabilities in the plasma is not fixed and for complete control, a steerable RF beam is required. The condition for RF absorption inside the plasma can be described as,

$$\omega_{\mathrm{RF}} = k_{\|} v_{\|} + \Omega(B(r)), \tag{3.1}$$

where, $\Omega(B(r))$ is the electron cyclotron frequency inside the plasma. In small fusion plasma experiments, the steerable mirrors have been used to control RF beam direction, which modifies the Doppler term ($k_{\|} v_{\|}$) to control the resonance position. The plasma instability control using steerable mirrors have been successfully demonstrated in [WSK+16]. In such cases, the movable parts are very close to the plasma surface and also require a necessary flexible launcher cooling system. Due to high thermal and neutron fluxes, this method is not preferable for future fusion power plants. This issue can be addressed by using remote steering launchers [PEK+10]. Alternatively, using a frequency tunable ECCD source, the localized energy transfer can therefore be possible by employing fixed launching mirrors.

Due to the large plasma volume in advanced magnetic confinement fusion experiments and the future fusion reactors, the magnetic field varies significantly from the plasma center to its edge. Consequently, the resonance condition between cycling electrons in a plasma and RF waves of a certain frequency is fulfilled in a small plasma layer. Therefore, it is desirable to have fast frequency step-tunable gyrotrons as ECCD sources, which allow to control the plasma instabilities employing fixed launching mirrors [ZT05] [TAB+01]. The frequency step-tunable operation of high power fusion gyrotron has been successfully demonstrated in [TAB+01], [GSA+14] as well as for the low-power spectroscopy gyrotrons in [BSC16], [CIO+09]. In the following sections, the possible approaches for the fast frequency step-tunability are studied for a 236 GHz DEMO gyrotron.

3.1.2 Selection of suitable modes

As discussed in section 2.1, the co-rotating $TE_{-43,15}$ mode has been selected for the operation at 236 GHz. As a next step, the new operating modes need to be selected to support fast-frequency tunability requirements of DEMO. The suitable mode selection criteria for the fast frequency step-tunability are listed below:

- The caustic radii of the selected operating modes has to be as close as possible, allowing optimum performance of quasi-optical launcher and mirror system for all the selected frequencies.
- As compared to the possible competitive modes, the coupling of the selected modes should be high.
- For the selected interaction section region, the mode eigenvalues must correspond to frequencies in the desired frequency band.

Based on the above criteria, the modes selected for frequency step-tuning within the range of -10 GHz to +10 GHz around 236 GHz (mode $TE_{-43,15}$) are enlisted in Table 3.1 with their corresponding operating frequency, relative caustic radii and beam radii. With these selected modes, the maximum

Table 3.1: Selected operating modes and their properties for fast-frequency step-tunable hollow-cavity gyrotron with frequency steps of 2 - 3 GHz.

f (GHz)	227.4	230.3	233.1	236	238.9	241.8	243.9
Δf	2.9	2.8	2.9		2.9	2.9	2.0
Mode	$TE_{40,15}$	$TE_{41,15}$	$TE_{42,15}$	$TE_{43,15}$	$TE_{44,15}$	$TE_{45,15}$	$TE_{43,16}$
Rel. caustic radius (r_c)	0.407	0.412	0.411	0.416	0.421	0.425	0.403
Beam radius (mm)	8.77	8.87	8.97	9.06	9.16	9.26	8.78
Q	1359	1385	1413	1443	1474	1507	1532
Wall loading (kW/cm^2)	1.83	1.89	1.93	1.98	2.08	2.15	2.24

relative caustic radii deviation is only 3.5 % from the average value, which guarantee that the same launcher and mirror system design works for all modes with very small radial shift of the Gaussian RF output beams and minimum stray radiation. Apparently, with the selected operating modes, frequency tunability in steps of 2 - 3 GHz can be achieved in the case of the 236 GHz hollow-cavity gyrotron. Assuming a beam radius of 9.06 mm, the mode spectrum around $TE_{-43,15}$ mode is presented in Figure 3.1. All neighboring modes within the frequency range of 224 GHz to 260 GHz having a relative coupling higher than 0.35 (with respect to $TE_{-43,15}$) are considered and the selected operating modes for a frequency step-tunability are marked in Figure 3.1. At a frequency of 236 GHz and beam radius of 9.06 mm, the coupling of selected modes for step-tunability is comparable to the main mode coupling. It should be noted that the relative coupling changes with the beam radius, and for each mode, the optimum value for the beam radius is given in Table 3.1.

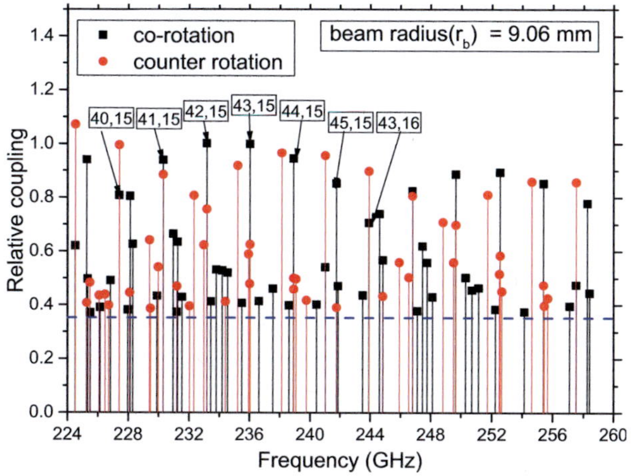

Figure 3.1: Mode spectrum in the cavity around the $TE_{-43,15}$ mode with electron beam radius of 9.06 mm. The selected modes for frequency tunability have been marked.

As discussed in section 2.1.1, the single-disk RF-window is transparent to the frequencies, which are multiple of the window basic frequency (see equation 2.1) and for the standard CVD-diamond window ($\varepsilon_r = 5.67$) (thickness 1.50 mm to 2.25 mm), the basic frequency is 42 GHz to 28 GHz. Thus, the single disk RF window does not support the fast frequency step-tunability of 2 - 3 GHz [TAH+01]. The alternative and very promising solution is to use a broadband Brewster angle window. As per the Brewster's law, the disk angle of $\Theta_B = \arctan\left(\sqrt{\varepsilon_2/\varepsilon_1}\right)$ is selected at which, the reflections of any wavelength is prevented. For the CVD-diamond, the value of Brewster angle is 67.2 °. Successful operation of the Brewster angle window has been demonstrated at KIT [AGM+12] [SDG+13] [GDJ+13]. The possibilities of gyrotron frequency step-tunability using a Brewster window have also been experimentally verified and reported in [LA90] [BDKT97] [GSA+14].

3.1.3 Frequency step-tuning: from high frequency to low frequency

The operating frequency of a gyrotron is close to the gyrofrequency of the electrons or its harmonics and can be calculated using equation 1.2. It is clear from equation 1.2 that the gyrotron operating frequency is highly dependent on the cavity magnetic field, while it is weakly dependent on the beam energy through the relativistic factor. Consequently, the magnetic field tuning is a very effective way to change the operating frequency of high-power gyrotrons. The effectiveness of magnetic field tuning is also experimentally validated in [GSA+14].

Based on the finalized operating parameters from section 2.4.2 for the operation at 236 GHz, the operating parameters for the fast-frequency step-tunable operation are selected by considering a constant beam current of 43 A and a minimum interaction efficiency of 36 % (without energy recovery). A pitch factor (α) of 1.25 is considered along with the realistic

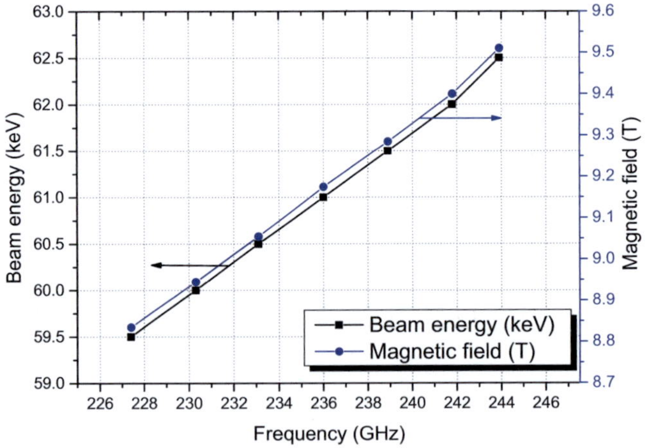

Figure 3.2: For a minimum interaction efficiency of 36 %, values of the beam energy and the magnetic field at different frequencies.

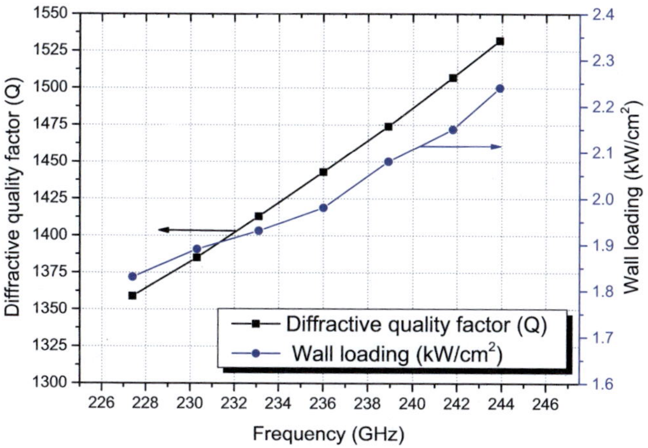

Figure 3.3: Cavity wall loading and diffractive quality factor at different operating frequencies.

Glidcop conductivity. The suggested values of the beam energy and the magnetic field for a particular frequency are plotted in Figure 3.2. The value of magnetic field sweep from 8.80 T to 9.48 T, which corresponds to the lowest (227.4 GHz) and highest (243.9 GHz) operating frequency. At each frequency step, the value of beam energy is also updated to maintain optimum detuning for efficient gyrotron operation. The dependency of the cavity wall-loading on the operating frequency is presented in Figure 3.3. The mode eigenvalue and quality factor increase with the operating frequency, while the relative caustic radius ($m/\chi_{m,n}$) is nearly constant for all the selected modes. According to the equation of wall-loading calculation for hollow cavity design (equation 1.27), these conditions lead to a rise in the wall loading with the operating frequency.

With the help of time-dependent, self-consistent, multi-mode simulations, the possibilities of fast-frequency step-tunability are theoretically verified and the result is presented in Figure 3.4. In this case, the operating frequency is tuned step-wise in the direction from high to low frequency. Over time, the variation of the cavity magnetic field and the beam energy is presented in first two plots, respectively.

As mentioned before, the beam energy is modified at every stage to maintain the optimized detuning for sufficiently high efficiency and output power. At each frequency step, the beam radius is selected for the maximum coupling (as per Table 3.1). The variation of the efficiency and output power is shown in the third and fourth plot respectively. The result supports the continuous frequency step-tunability from the high-frequency towards the low-frequency with minimum interaction efficiency of more than 35 % and stable output power close to 1 MW.

In the fusion plasma experiments, the growth rate of a possible plasma instability determines the speed of frequency step-tunability. It is in the order of several GHz/s for large scale fusion experiments [Thu11]. In gyrotron, the speed of frequency step-tunability is limited by the rate of the change of the operating magnetic field. Using auxiliary solenoids, the axial magnetic field is adjusted to achieve frequency step-tunability within limited frequency range, and the magnetic field from the main solenoid is kept constant. The magnet system design for a 236 GHz gyrotron is not finalized yet at IHM-KIT. Therefore, it is not possible to determine the speed of frequency-tunability at this stage.

3.1.4 Frequency step-tuning: from low frequency to high frequency

As discussed in the previous section, the magnetic field tuning is effective for the frequency step-tuning from high frequency to low frequency. However, due to hysteresis effects [DN10] [DII$^+$03], it is not possible to achieve frequency tunability at the desired 2 - 3 GHz steps from low frequency to

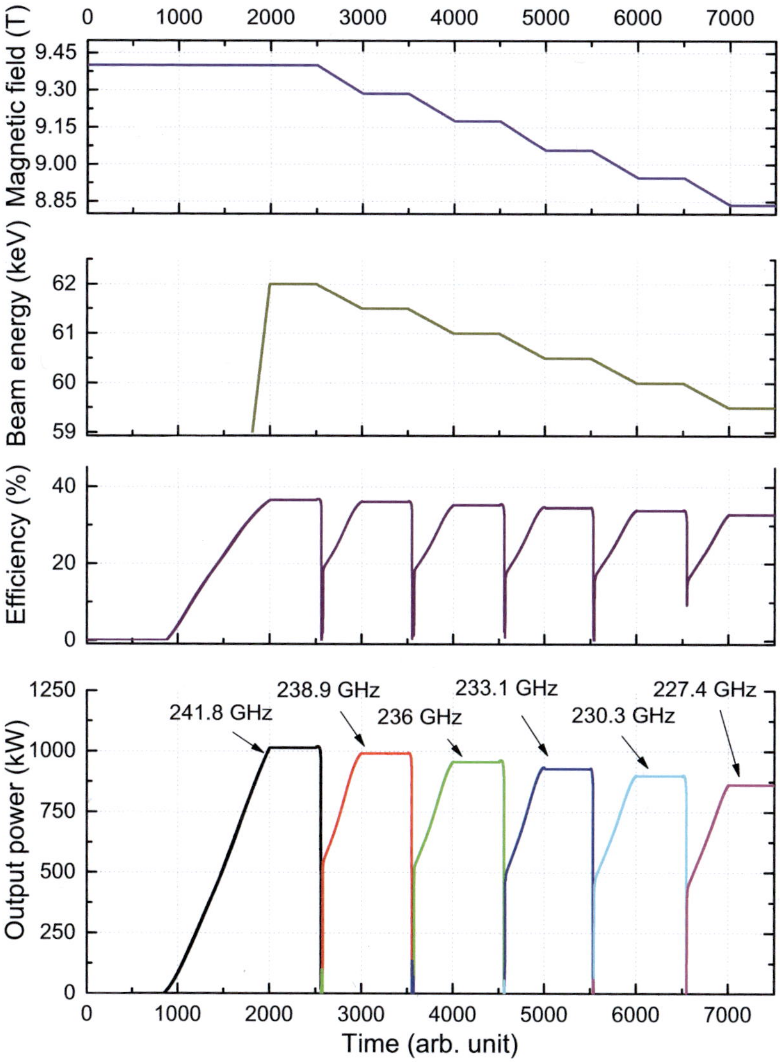

Figure 3.4: Step-wise fast-frequency tuning with a changing magnetic field and beam energy [KAF+16a].

high frequency by simply changing the magnetic field. The possible approaches of low frequency to high frequency step-tunability are verified in this section. This case has been studied considering frequency step-tunability from 236 GHz, $TE_{-43,15}$ mode to 238.9 GHz, $TE_{-44,15}$ mode.

Using single-mode, self-consistent calculations, the output power of the $TE_{-43,15}$ and $TE_{-44,15}$ modes has been calculated and presented in Figure 3.5. Due to critically high detuning, it is not possible to excite $TE_{-43,15}$ and $TE_{-44,15}$ modes with magnetic field less than 9.16 T and 9.28 T, respectively. The selected optimum magnetic field for the excitation of the modes $TE_{-43,15}$ and $TE_{-44,15}$ with high efficiency is 9.175 T and 9.285 T. To achieve low frequency to high frequency step-tunability simply with magnetic tuning, when the magnetic field (B_0) are increased from 9.175 T to 9.285 T, the $TE_{-43,15}$ mode remained excited with low detuning and low

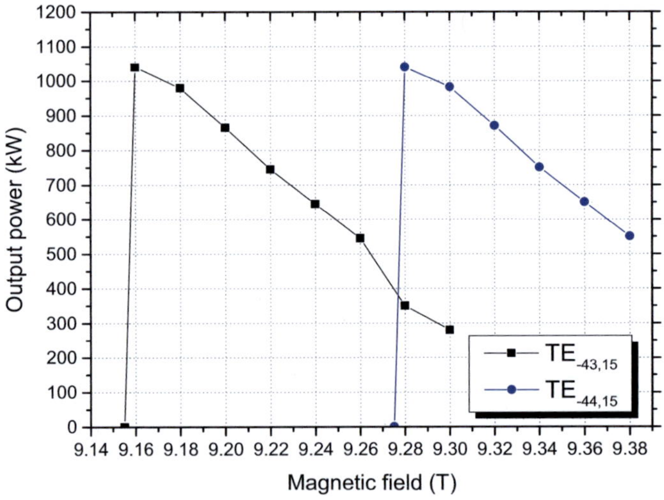

Figure 3.5: Output power of the modes $TE_{-43,15}$ and $TE_{-44,15}$ at different values of the magnetic field.

output power. This condition prevents the continuous frequency tuning from low frequency to a higher frequency. The similar gyrotron behavior has been also verified in [DN10] for a 170 GHz ITER gyrotron, considering main operating mode $TE_{-32,9}$ and next parasitic modes $TE_{-31,9}$ and $TE_{-33,9}$.

The alternative approach for frequency step-tunability from low frequency to high frequency is presented in Figure 3.6. As the gyrotron cavity is compatible with all selected frequencies, the individual mode can be excited separately. In this approach, the beam energy is reduced to a low level until the working mode is detained and then excite the desired mode at the higher frequency by increasing the beam energy again. The magnetic field is also modified accordingly. For example, in Figure 3.6, initially $TE_{-43,15}$ mode is excited at frequency of 236 GHz (till t = 2500). The beam energy is increased from 30 keV (t = 0) to 61 keV (t = 2000) and the constant magnetic field of 9.175 T is considered. To excite $TE_{-44,15}$ mode at a frequency of 238.9 GHz, the beam energy is reduced to the 20 keV which, suppress the excitation of $TE_{-43,15}$ mode (t = 2500 to 3000). From t = 3000 to 3500,

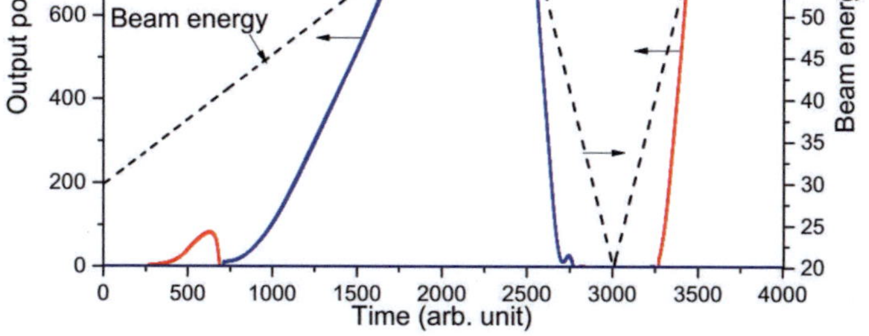

Figure 3.6: Possible approach for low-frequency to high-frequency tuning.

the beam energy is again increased till 61.5 keV with the updated magnetic field of 9.285 T and the desired mode $TE_{-44,15}$ is excited at a frequency of 238.9 GHz. Similar to the high frequency to low frequency tuning, the speed of the frequency tuning from low frequency to high frequency tuning is dependent on the rate of change of the cavity magnetic field.

3.2 Multi-purpose operation

The development of a DEMO EC system is focused to provide maximum current drive efficiency and it is the demonstrated for the pulsed and steady state operation [PTZ+12]. In addition to the current drive (ECCD) application, the possibilities of plasma heating are also required at lower operating frequencies [TFK+15]. For DEMO (EUROFusion baseline 2012, aspect ratio (A = 4)), the desired gyrotron frequency, output power and operation mode (heating (H) or current drive (CD)) for the different gyrotron applications (e.g. Start-up, ramp up, bulk heating and plasma instability control) are estimated in [GBB+15]. The ratio of the frequency selected for plasma heating application (f_1) to the suitable frequency for the current drive application (f_2) is approximately 0.8 - 0.833 for DEMO [GBB+15]. The frequency of the heating mode (f_1) is only dependent on the toroidal magnetic field, while the selection of current drive frequency (f_2) is dependent on the electron temperature, the launching angle and the toroidal magnetic field.

To fulfill these requirements of multi-purpose operation, the multi-frequency gyrotron is a prerequisite. The requirement of multi-purpose gyrotrons for future fusion power plants are also suggested in [STK+06c]. The multi-purpose operation of high power gyrotrons has been successfully demonstrated in [SKO+12] [SKO+13] [IOK+16] (104 GHz / 137 GHz / 170 GHz / 203 GHz), [KYS+15] (110 GHz / 138 GHz), [WSK+16] (105 GHz / 140 GHz) and [DLA+08] etc. In this section, feasibility of the multi-purpose operation of a 236 GHz DEMO gyrotron is investigated and suitable operating parameters are suggested. The selection of the suitable frequencies for the multi-frequency operation and corresponding operating modes has been already discussed with details in section 2.1. For the single disk CVD-diamond window (ε_r = 5.67) with disk-thickness of 1.861 mm, the basic frequency of an RF-window is around 33 GHz. Based on it, the operating frequencies of 170 GHz, 203 GHz, 236 GHz and 269 GHz have been selected. The window thickness is an integer multiple of $\lambda/2$. The selected window is transparent to these frequencies, which allow lossless transmission of the wave. The possible applications of the selected frequencies at various tokamak aspect ratio (A) have been presented in Table 2.1. The selection of corresponding operating modes for multi-purpose operation is based on the launcher design. The relative caustic radius of selected modes is identical to ensure efficient mode conversion using internal launcher and position of output beam at the center of the window.

After finalizing the operating modes, the suitable operating parameters are selected for the multi-frequency operation. The physical parameters of the cavity are enlisted in Table 2.4. For each case, operating parameters are finalized considering the wall-loading limitation of 2 kW/cm^2 and efficient gyrotron operation. The velocity ratio (α) of 1.25 is considered for all four cases. The frequency dependent Glidcop conductivity is calculated using equation A.5. The final operating parameters are presented in Table 3.2. As discussed in section 3.1.3, cavity wall-loading limits the maximum input power at high operating frequencies which, eventually reduce the output power. For each case, the stability of operation is verified using realistic time-dependent, multi-mode simulations. The setup of the simulations is similar to those presented in section 2.3.2. The standard diode type start-up scenario has been considered for the simulations. Using suggested cavity design and operating parameters, the stable operation is possible for all frequencies with the controlled mode competition. Considering the limitations of the magnet system, the operating frequency can be tuned in slow-time scale (within a few minutes). The results are also verified using the realistic beam parameters with a velocity spread of 6 % and the radial width of $\lambda/4$.

3.3 Launcher design for step-tunable and multi-frequency operation

A quasi-optical mode converter transforms the high order operating modes into a fundamental Gaussian beam and also separates an RF-beam from the

Table 3.2: Multi-frequency operation of the DEMO gyrotron using realistic Glidcop conductivity.

Frequency (GHz)	170.0	203.0	236.1	269.1
Mode	$TE_{-31,11}$	$TE_{-37,13}$	$TE_{-43,15}$	$TE_{-49,17}$
Mode eigenvalue	74.325	88.769	103.213	117.656
Relative caustic radius	0.4171	0.4168	0.4166	0.4165
Magnetic field (T)	6.785	7.975	9.165	10.349
Beam radius (mm)	9.13	9.10	9.06	9.04
Beam electron energy (keV)	81	70	61	55
Beam current (A)	59	48	43	38
Quality factor (Q_{diff})	820	1171	1443	1839
Ohmic wall loading (kW/cm^2)	2.00	1.99	2.00	1.99
Conductivity ($\times 10^7$ S/m)	2.1	2.0	1.9	1.8
Without velocity spread or radial spread consideration				
Output power (kW)	1650	1220	960	821
Interaction efficiency (%)	35	37	38	40
With 6 % velocity spread and $\lambda/4$ radial width				
Output power (kW)	1550	1150	920	765
Interaction efficiency (%)	33	35	36	38

spent electron beam. It consists of a quasi-optical launcher (QOL) and a mirror system. The fundamentals of QOL have already been discussed in section 1.3.4. The launcher surface is deformed to convert high order TE modes to Gaussian like distribution. In this section, the first design of a hybrid-type launcher is suggested for a 236 GHz hollow-cavity gyrotron and its performance is verified for fast-frequency step-tunability and multi-purpose operation.

The harmonically deformed launcher is effective only for cavity modes having a relative caustic radius close to 0.5 [JTP+08]. In the case of the 236 GHz, $TE_{-43,15}$ mode DEMO gyrotron, the relative caustic radius is 0.416. Therefore, the harmonically deformed launcher is not a good choice for high Gaussian mode content. Due to the limitations of a relative caustic radius, either mirror-line or hybrid-type launcher has to be selected. The performance comparison of the mirror-line launcher and the hybrid-type launcher is presented in [JTGJ16] and [JGJ+15]. Compared to the harmonically deformed launcher and mirror-line launcher, the hybrid-type launcher provides unique advantages such as, high Gaussian mode content, low diffraction losses and a small range of surface perturbation. The detailed synthesis and optimization method for the hybrid-type launcher are presented in [JTGJ16].

The in-house code "TWLDO" is used for the launcher synthesis and analysis [JTGJ16]. The field distribution on the launcher wall of the final design is shown in Figure 3.7. The design goals of the launcher is to achieve maximum Gaussian mode content with minimum diffractive losses at the cut of the launcher. The input radius of the launcher is 22.37 mm (corresponding to $1.07 \cdot r_{cav}$). The launcher length and cut length of the launcher design are 227.11 mm and 50 mm, respectively.

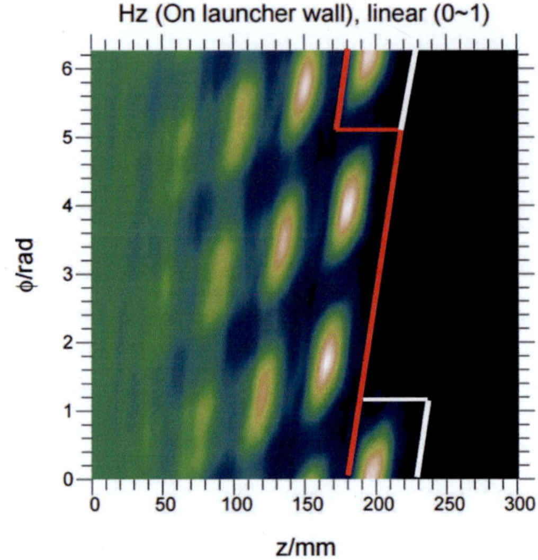

Figure 3.7: Field distribution on the hybrid-type launcher wall for the 236 GHz $TE_{-43,15}$ mode gyrotron. The edges of the cut and launcher aperture are indicated in red and white line, respectively.

Table 3.3: The Fundamental Gaussian Mode Content (FGMC) of the designed launcher for the selected operating frequencies the multi-purpose multi-frequency gyrotron operation.

Frequency (GHz)	170.0	203.0	236.1	269.1
Cavity TE mode	$TE_{-31,11}$	$TE_{-37,13}$	$TE_{-43,15}$	$TE_{-49,17}$
Rel. caustic radius	0.4171	0.4168	0.4166	0.4165
FGMC (%)	98.6	98.2	97.4	96.0

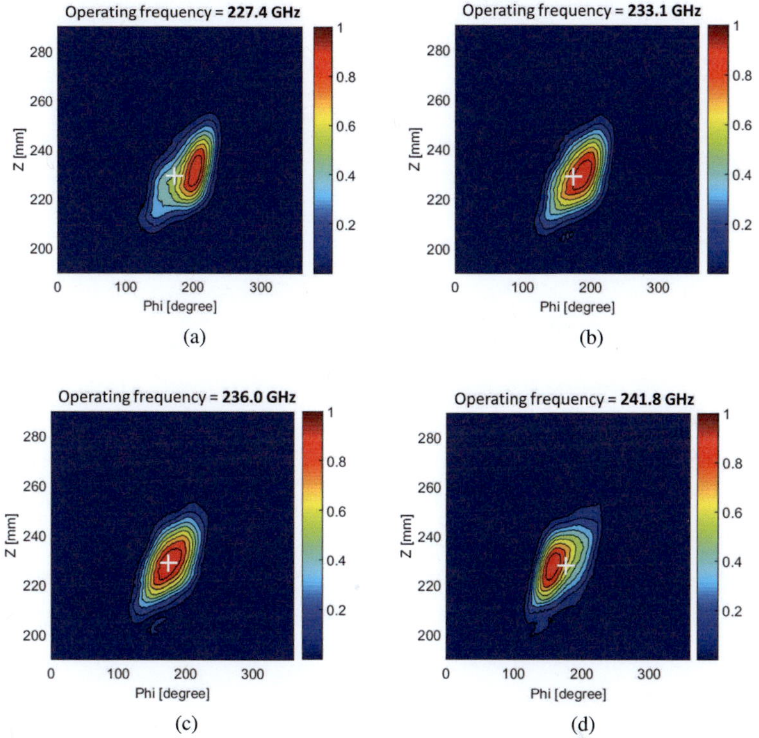

Figure 3.8: Simulated radiation pattern of the designed launcher at the radial distance of 60 mm using the SURF3D code package for the operating frequencies of (a) 227.4 GHz (b) 233.1 GHz (c) 236.0 GHz and (d) 241.8 GHz.

The performance of the designed launcher is simulated for a multi-frequency operation and the results are presented in Table 3.3. More than 96 % of Fundamental Gaussian Mode Content (FGMC) has been achieved with all selected frequencies. As the relative caustic radii of all selected modes for a step-frequency tunability are also close to each other, the performance of the launcher is nearly identical for all selected modes at various frequencies. The simulated FGMC for the different frequencies are listed in Table 3.4.

Table 3.4: Performance of the designed launcher at various frequencies of the step-frequency tunable gyrotron.

Frequency (GHz)	227.4	230.3	233.1	236.1	238.9	241.8	243.9
Cavity TE mode (indices m,n)	40,15	41,15	42,15	43,15	44,15	45,15	43,16
Rel. caustic radius	0.402	0.407	0.412	0.417	0.421	0.427	0.403
FGMC (%)	93.11	95.42	96.67	97.31	97.02	95.88	92.35

Using the commercial 3-D full-wave code SURF3D [Nei04], the radiation pattern of the launcher is also calculated for all frequencies at a radial distance of 60 mm. Figure 3.8, the radiation pattern at operating frequencies of 227.4 GHz, 233.1 GHz, 236 GHz and 241.8 GHz is presented. Compared to the field profile at the frequency of 236 GHz, the RF beam is slightly shifted for the other frequencies due to the small difference in their relative caustic radii, however, the overall beam profile remains Gaussian-like with the the average FGMC of more than 95 %.

4 Beam misalignment tolerance and space-charge neutralization

In this chapter, the gyrotron operation is investigated considering the effects of beam misalignment and space-charge neutralization. In the first part, the influence of axial beam misalignment is studied for a 236 GHz hollow-cavity DEMO gyrotron including a velocity spread and a radial width of beam. In the second part of the chapter, the effect of space-charge neutralization on the beam parameters is studied, followed by performance analysis of the DEMO gyrotron in long-pulse operation.

4.1 Study on beam misalignment tolerance

4.1.1 Motivation for the beam misalignment analysis

In an ideal scenario, the gyrotron electron beam should be located concentric around the cavity axis, but due to a misalignment of super-conducting magnets with gyrotron and due to mechanical tolerance of gun, an off-axis electron beam is generated. Because of the very short wavelength and dense mode spectrum, the performance of the high frequency gyrotron is greatly influenced by the beam misalignment. The interaction efficiency is also reduced due to reduction in mode-beam coupling. In addition to that, beam misalignment also affects the start-up scenario and mode competition because of the changes in the overall coupling conditions.

Additionally, the cavity wall-loading is increased due to the electron beam misalignment, which is one of the critical issues for the high power (several kW to few MW) fusion gyrotrons [ZZMS11]. In the case of THz gyrotrons, the beam misalignment causes significant effects on gyrotron performance because of the small cavity radius [ISN+98]. The quasi-optical launcher and mirror systems are designed only for operating modes in one particular rotation. The co-rotating mode is preferred in most cases due to relatively high beam coupling. The electron beam misalignment also increases the coupling of the counter-rotating modes, and it is important to know the allowable electron beam misalignment for the mode excitation in a single rotation.

The effect of beam misalignment has been experimentally studied for various gyrotrons and the results are discussed in [SDG+13] (140 GHz, $TE_{-22,8}$), [ISN+98] (335 GHz, $TE_{-3,3}$), [BKT94](146 GHz, $TE_{-16,2}$) and [ZZMS11]. These experiments suggest the stable operation of the main mode till axial beam misalignment of $\lambda/4$ - $\lambda/3$. Various numerical models have been suggested to analyze the influence of off-axis electron beam. To study an influence of the beam misalignment on gyrotron efficiency, the analytic model based on the beam-mode coupling has been proposed in [LN84]. Using fixed-field approximations and starting-current calculations, the effects of eccentric beam has been estimated in [DS92], [Dum94] and [Zha99]. Few important theories to estimate performance of gyrotron with off-axis electron beam have been presented in [NDL95] [Air03] [KDNI14].

In this work, the effect of beam misalignment on gyrotron performance is studied using macro-electron beam with realistic beam parameters. In the case of high-frequency fusion gyrotrons, this method provides unique advantages such as:

- The realistic electron beam parameters can be included in the misalignment study. As discussed in 2.3.3, performance of the high-frequency, high-power gyrotrons are highly dependent on the realistic beam parameters. The mode-beam coupling is varied according to the position of the beam and its thickness. Specially for the high-frequency DEMO gyrotrons (frequency > 200 GHz), inclusion of realistic beam parameters (radial width, velocity spread etc.) is a prerequisite for the accurate assessment of the effects of beam misalignment.

- The mode spectrum density also increases significantly with high-order operating modes. This method has possibilities to include effects of sufficiently large numbers of neighboring modes to determine the performance of the gyrotron.

- The variation of axial EM-field and beam parameters during the operation shall be considered for analysis, which suggests the use of non-linear, self-consistent gyrotron theories.

- For the calculation of particle bunching and trajectories, the modified beam coupling with off-axis electron beam can be included in the simulations.

- Due to the practical tolerances of the gun technologies, it is difficult to get the complete homogeneous electron beam. The effects of inhomogeneous beam have been studied in [ZIP+15]. These effects can be also considered in the macro-electron beam based beam-misalignment study.

For the 170 GHz gyrotron, the results based on a misaligned macro-electron beam approach are presented in [APIT13]. In which as a preliminary step, the realistic misaligned electron beam is generated using three-dimensional, electrostatic, self-consistent code ARIADNE [PV04] and used as an input for multi-mode, time-dependent simulation. In the case of a 236 GHz DEMO gyrotron, the magnetic system and MIG design are not finalized yet, therefore, the approach suggested in [APIT13] is not suitable at this development stage.

To generate the realistic misaligned electron beam, new code-package named "VISTAR" has been developed, which can generate misaligned macro-electron beam with desired beam properties. The detailed features of this code are presented in section A.4 and the used velocity spread and radial width implementation models are discussed in section A.3. Using this code-package, the requirement of gun design for beam misalignment tolerance analysis is eliminated.

4.1.2 Setup of the electron beam in the misalignment study

For the accurate gyrotron interaction simulation using the misaligned electron beam, the macro-electron beam with realistic beam parameters is generated and is further used for the start-up simulations. For the particular start-up scenario, the properties of macro-electrons are assigned in a separate file and used as an input for interaction simulation. The user-defined properties of macro-electron include electron energy (V_{b}), relativistic factor (γ), velocity ratio (α), guiding center radius (r_{gc}), guiding center phase (ϕ_{gc}) and phase (ϕ). The detailed macro-electron implementation is presented in Figure 4.1. The cavity azimuthal space is divided into equal number of sections and the each section is assigned to a particular ϕ_{gc}. In Figure 4.1, the azimuthal space is divided into 50 parts and 50 different ϕ_{gc} are selected.

Based on the algorithm discussed in section A.3.3, various guiding center radii (r_{gc}) are selected for each guiding center phases (ϕ_{gc}). In Figure 4.1, ten different guiding center radii (r_{gc}) are selected for each ϕ_{gc}. Using the velocity spread implementation method as discussed in section A.3.1 and A.3.2, the values of velocity ratio have been selected and assigned to particular electrons (see Figure 4.1). At the end, the number of phases (ϕ) is also selected and different phase values are assigned to each electron. The phase of each electron is linearly separated by the value of $(2 \cdot \pi)/n$, where n is the selected number of phases.

It is clear that each electron has unique parameters, which are in different combination of guiding center phase (ϕ_{gc}), guiding center radius (r_{gc}), velocity ratio (α) and phase (ϕ). For easy understanding the electron shall be labelled as (ϕ_{gcw}, r_{gcx}, α_y, ϕ_z) and the values of w, x, y and z for this case are 50, 10, 20 and 9 respectively. The total number of electrons required at particular beam energy is simply the multiplication of the selected different values of each physical parameters. The generated macro-electron beam is used as an input for the multi-mode start-up simulations. Considering diode-type start-up scenario, the beam energy is increased stepwise, while

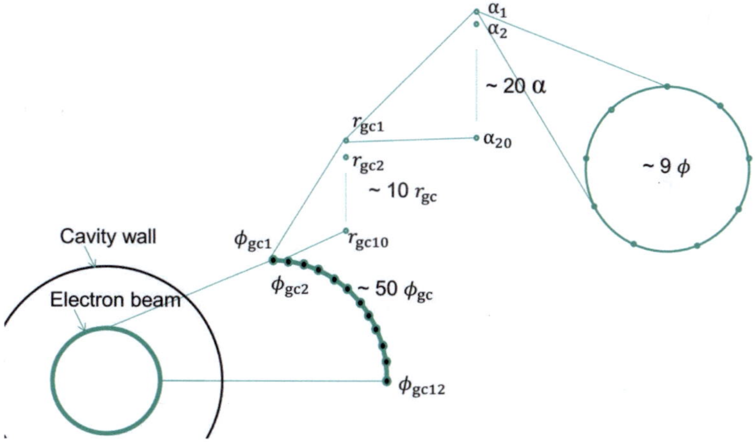

Figure 4.1: Macro-electron beam for multi-mode time-dependent simulation. For this example, the total azimuthal space of the cavity is divided into 50 parts. Ten different guiding centers are used to generate a radial width of the beam. Twenty velocity samples are considered to implement a perpendicular velocity spread with nine different phase values.

the beam current and the velocity ratio changes adiabatically. The values of the macro-electron beam are updated with each beam energy step.

Effect of a beam misalignment on the relativistic factor

In case of an aligned ideal electron beam (without radial width), the voltage depression for the hollow cavity gyrotron can be calculated using [DK81],

$$\Delta V_{\text{hollow}} = \frac{I_{\text{b}}}{2\pi\varepsilon_0 c\beta_{\|}} \cdot \ln\left(\frac{r_b}{r_{cav}}\right) \tag{4.1}$$

For the thick electron beam, a spread in the voltage depression is inevitable and an axial beam misalignment further add to it. Unlike the ideal condition, in the realistic electron beam, each electron experiences different potential difference which results in a spread of voltage depression. This spread in the voltage depression can be easily translated to the equivalent energy spread, which is critical for stability of the operating modes. Along with the radial width and beam misalignment, non-uniform emission of emitter also leads to the beam energy spread. The effect of non-uniform emission on gyrotron performance has been systematically studied in [ZIP+17] [Zha16]. In this section, the influence of a radial width and an axial misalignment on the beam potential are studied.

Based on the theories of mirror charges, the in-house code "WickedQueen" has been developed at IHM-KIT [FAG+16]. Considering an electron beam with 20000 uniformly distributed electrons, the influence of an axial misalignment on the voltage depression is calculated for a 236 GHz hollow-cavity design. The electron beam width of $\lambda/4$ is selected for the analysis. The spread in the voltage depression is calculated for the beam misalignment (D/λ) between 0 to 0.9 mm in steps of 0.1 mm. The results of this analysis are presented in Figure 4.2. The similar simulations are also performed for a 238 GHz coaxial cavity gyrotron and the results are compared

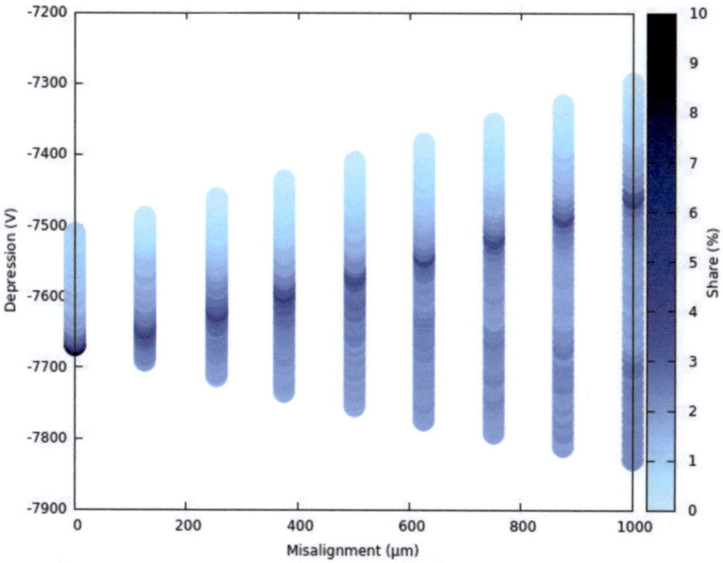

Figure 4.2: Effects of off-axis electron beam on the voltage depression of a 236 GHz hollow-cavity gyrotron [IKF+17].

in [IKF+17]. The interval of observed voltage depression values starts at 200 V and grows linearly with increasing misalignment at the rate of 0.3 V/μm for the hollow cavity, which is equivalent to the 2 V/μm for coaxial cavity design. From the calculated spread in the voltage depression, the equivalent energy spread of the misaligned beam can be calculated using equation 1.3.

For the hollow-cavity design, the influence of an axial misalignment on the voltage depression and equivalent energy spread is significantly low. For example, in the case of an axial misalignment of 0.381 mm ($D/\lambda = 0.3$), the equivalent spread in the relativistic factor (γ) is only 0.005 %. Therefore, for a 236 GHz hollow-cavity design, the effects of energy spread are not considered for the beam misalignment study.

4.1.3 Study of the beam misalignment using a macro-electron beam

In this section, the simulation setup for the beam misalignment study is briefly discussed. The complete method is divided into two steps. Initially, the misaligned beam is prepared using macro-electrons. The prepared electron beam file is further used as an input for the time-dependent, self-consistent simulations and performance of the gyrotron is determined for the particular misalignment.

Step 1: Preparing an off-axis beam file using realistic beam parameters

In case of an aligned electron beam, the average particle guiding center radii are equal to the beam radius, while for the off-axis macro electron beam, for particular misalignment (D), the average guiding center radii are calculated using,

$$r_b^{'}(\phi_{gc}) = \sqrt{r_b^2 + D^2 + 2 \cdot r_b \cdot D \cdot \cos\phi_{gc}}. \tag{4.2}$$

Here, for each guiding center phase ϕ_{gc}, $r_b^{'}(\phi_{gc})$ is the modified average beam radius with the axial beam misalignment of D. First, the misaligned electron beam is generated without a radial width and a velocity spread for the particular value of an axial-misalignment. In the next step, the desired realistic beam properties are introduced into the ideal misaligned beam. The updated average guiding center radii from equation 4.2 are selected and the required radial width is implemented around these values. As per the simulation setup, either velocity spread or radial width or both properties are introduced into the ideal misaligned beam. The numerical approach presented in section A.3.1 and A.3.2 is used for velocity spread implementation and used radial width implementation approaches are discussed in section A.3.3.

For the simulations using a macro-electron beam, the selection of realistic beam parameters (e.g. a number of different values for velocity spread, radial width, guiding center phases etc.) is critical and has great impact on the computational resources. For example, selecting 50 values for guiding center phases, 50 values for the velocity spread, 10 distinct values for radial width and 9 different phase leads to the 225000 numbers of particles per energy step (refer Figure 4.1). So, the careful selection of the simulation parameters is necessary to obtain the required accuracy with the optimal resources.

Step 2: Multi-mode simulation using a macro-electron beam

The macro-electron beam file generated from step 1 is used as input for the multi-mode, time-dependent, self-consistent simulation. In the EURIDICE code package, it is possible to import an external electron beam file for interaction simulation.

4.1.4 Results of the misalignment study

The behavior of a 236 GHz gyrotron is verified for different values of the beam misalignment using the discussed macro-electron based approach. The multi-mode, time-dependent simulations are performed considering an axial-beam misalignment (D/λ) of 0.1, 0.2, 0.25 and 0.3, which corresponds to 0.127 mm, 0.254 mm, 0.317 mm and 0.381 mm assuming a wavelength of 1.2711 mm (236 GHz). The macro-electron beams are generated with 50 uniformly divided guiding center phases (ϕ_{gc}). The physical parameters of the selected cavity design are presented in Table 2.4 (L_1 / L_2 / L_3 = 16 mm / 12 mm / 16 mm, θ_1 / θ_3 = 2.5 ° / 2.0 °, D_1 / D_2 = 4 mm / 5 mm). The finalized gyrotron operating parameters from section 2.3.4 are selected for this analysis (V_b = 58 keV, I_b = 39 A, B_0 = 9.145 T).

Table 4.1: Effect of beam misalignment on gyrotron operation (radial width = $\lambda/4$, no velocity spread).

Beam misalignment (mm)	Beam misalignment (D/λ)	Output power (kW)	Interaction efficiency (%)
0	0	760	35
0.127	0.1	748	34.5
0.254	0.2	733	33.5
0.317	0.25	720	33
0.381	0.3	unstable mode	

Initially, for the first estimation, only a radial width without a velocity spread of the electron beam is considered to the control computational time and resources. For particular value of a guiding center spread, 10 distinct values of guiding center radius are selected to implement $\lambda/4$ radial width, along with 9 phases (ϕ) values. The simulations are performed using Helios IFERC-CSC supercomputer facility and in total 992 numbers of processors was used for the simulations. Considering 32 neighboring modes, the time required for each simulation is around 19.5 hours.

The results from the multi-mode self-consistent simulations are summarized in Table 4.1 and one such result with beam misalignment of 0.317 mm is presented in Figure 4.3. The beam energy is increased from 28 keV to 58 keV in the steps of 0.5 keV. The result suggests a stable gyrotron operation till an electron beam misalignment of 0.317 mm (D/λ = 0.25). With an axial electron beam misalignment of up to 0.317 mm, the desired operating mode $TE_{-43,15}$ is excited successfully, but due to the reduction of main mode coupling, the output power and efficiency is decreased with an increase in the beam misalignment.

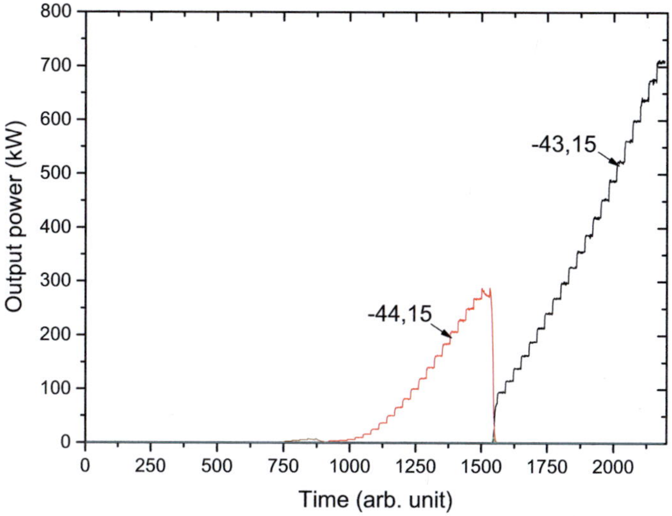

Figure 4.3: RF behavior of a 236 GHz, $TE_{-43,15}$ mode gyrotron with axial beam misalignment of 317 μm. An electron beam with a radial width of $\lambda/4$ is considered for the analysis.

As a next step, the effect of a velocity spread is included together with the assumption of a radial width. The equal area method (section A.3.1) is used for the Gaussian-type velocity spread implementation. The beam energy is increased from 28 keV to 58 keV with the steps of 0.25 keV. To maintain sufficient accuracy, 50 distinct values are used for the velocity spread implementation (6 % (rms)), along with 50 guiding center phase, 20 values for radial width and 9 phases. The total number of particles per beam energy has increased to 450,000, which significantly increases the computational resources (1600 processors, around 66 hours). For the axial-beam misalignment of 0.3 mm, the result of a multi-mode, time-dependent analysis, including velocity spread is presented in Figure 4.4. The stable excitation of the operating mode $TE_{-43,15}$ is possible until an axial misalignment of 0.3 mm ($D/\lambda = 0.23$).

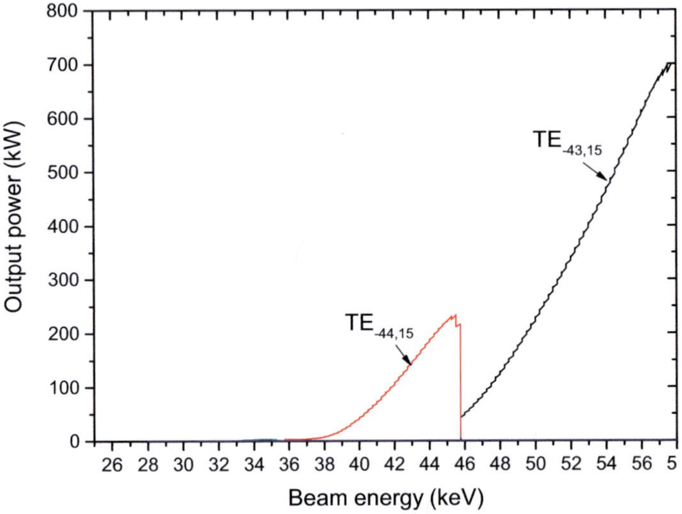

Figure 4.4: The multi-mode start-up simulations with an axial beam misalignment of 0.3 mm. The effects of electron velocity spread and radial width have been also included. For precise estimation, a macro-electron beam with a large number of electrons has been used with the beam increment step of 0.25 keV.

For a 236 GHz DEMO gyrotron, the beam misalignment tolerance suggested by this method ($D/\lambda = 0.23$, D = 0.3 mm) is in good agreement with the experimental results presented in [SDG+13] and recently in [YST+17]. In case of the W7-X gyrotron, the beam misalignment tolerance is in order of 0.5 mm – 0.6 mm ($D/\lambda = 0.33 - 0.25$). The similar value of tolerances is also suggested in [ZZMS11]. It should be noted that for a high frequency gyrotrons, beam misalignment tolerance is very critical and precise beam placement is a prerequisite for the stable gyrotron operation.

4.2 Effect of space-charge neutralization

In the ideal gyrotron operation, a perfect vacuum condition is considered inside the device, but because of the practical limitations, the residual gas is still present at very low pressure of $10^{-8} - 10^{-10}$ mbar [Pio90]. Even under such good vacuum condition, the residual gas is ionized by the collisions between the energetic beam electrons and gas molecules (e.g. $N_2 + e \rightarrow N_2^+ + e + e$). For secondary electrons, which are produced simultaneously with the ionization of the residual gas, the space charge potential is repulsive, therefore, they diffuse rapidly back to the beam and transported to the gun or the collector wall. The positive ions formed in a potential trap can only leave the potential trap: (a) adequately by collisions with other particles gain energy (diffusion in velocity space) in order to overcome the potential barriers or (b) by recombination with electrons (for example: $N_2^+ + e \rightarrow N + N$) and form neutral atoms or molecules.

The positive ions accumulate in the potential-well along the gyrotron axis and the beam energy rise uncontrollably. This may change the resonance conditions and the effective operating parameters significantly, specially during the continuous wave (CW) operation. So, for CW gyrotrons, it is important to know the effects of space-charge neutralization and the behavior of positive ions inside the gyrotron. The possible effects of space-charge along the beam are listed here.

- The positive space-charge accumulates in the potential trap and increases the beam energy at the resonator.
- A neutralizing plasma between the cathode and the resonator can lead to beam oscillations and modifies the beam properties.
- Lowering potential along the beam path may restrict the operating range of parameters and limit the maximum achievable velocity ratio (α).

The various studies on the space-charge neutralization during the CW gyrotron operation have been performed. The frequency tunning caused by neutralization has been investigated in [Sch15] and the analytic model

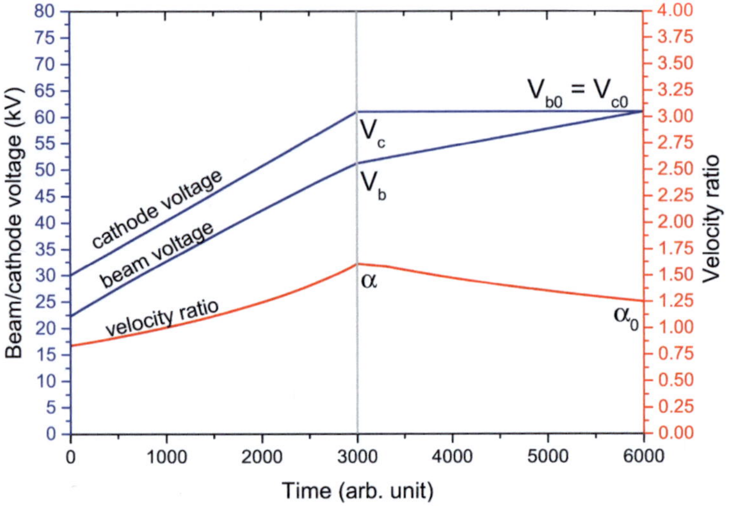

Figure 4.5: Variation of the beam parameters during the neutralization process.

for ion compensation calculation has been presented [MS16]. Using full-gyrotron simulations, the neutralization process has been studied in [PAI+14]. In this work, the effects of space-charge neutralization on electron beam parameters is presented, followed by the multi-mode start-up simulations considering the updated beam parameters.

4.2.1 Variation of beam parameters during the long-pulse gyrotron operation

The variations in the beam parameters due to space-charge neutralization are studied in this section. For the full beam neutralization scenario, the variation in the beam parameters (e.g. beam energy, cathode voltage, pitch factor etc.) is presented in Figure 4.5 for a 236 GHz DEMO gyrotron. The complete gyrotron start-up scenario can be divided into two stages and both stages are separated by a vertical line. The corresponding operating parameters have been enlisted in the Table 2.8.

Stage I: start-up phase

In this phase, the cathode voltage increases linearly till its optimum value. Due to voltage depression, the beam energy is lower than the cathode voltage and increases with the cathode voltage. The beam energy and the velocity ratio before neutralization are denoted as V_b and α, respectively. The change in the pitch factor ($\alpha = \beta_\perp/\beta_\|$) can be estimated using $\beta_\perp = \gamma_0 \cdot \beta_{\perp 0}/\gamma_{\rm eff}$. Here, γ_0 and $\beta_{\perp 0}$ are the values of relativistic factor and normalized perpendicular velocity without voltage depression, respectively. $\gamma_{\rm eff}$ is the value of the relativistic factor before the space-charge neutralization. In stage I, as the beam energy is less than its nominal value (61 keV), the value of the pitch factor is higher than the desired value of 1.25. The beam current also increases adiabatically, however, it is not plotted in Figure 4.5.

Stage II: neutralization phase

Due to the ionization of residual gas molecules, positive ions accumulate in the potential wells and compensate the depression voltage. In the case of 100 % neutralization, beam energy rises up to the cathode voltage and accordingly the pitch factor drops to the optimal value. The beam energy and velocity ratio after neutralization are indicated as V_{b0} and α_0, respectively.

In case of hollow-cavity gyrotrons, instead of full beam neutralization, 60 % neutralization has been suggested in [Sch15] [PAI+14]. The variation of beam parameters with 60 % neutralization is presented in Figure 4.6. Using EURIDICE code-package, the multi-mode start-up simulation shall be performed in two continuous parts, and equivalent simulation environment can be created. In the first stage, the cathode voltage is controlled and raised to the optimum value, while in the second stage, the beam energy is modulated as per the change in the voltage depression.

4.2.2 Multi-mode, self-consistent simulations including space-charge neutralization effect

For the multi-mode, time-dependent simulations with space-charge neutralization effect, it is necessary to know the precise value of the pitch factor (α) at the end of the start-up phase (stage I). The value of pitch factor without beam neutralization (α) can be selected either from gyrotron experimental results, or estimated from previous experiences. For example, in the case of 140 GHz W7-X gyrotron, the value of pitch factor before neutralization (α) is estimated to 1.45 ($= \alpha_0 + 0.2$) for the voltage depression of $\sim$ 5 kV. For the high-frequency gyrotrons with very high order operating modes (eigenvalue > 100), the value of voltage depression is more than 7 kV and alternative method is required to precisely determine the beam parameters before space-charge neutralization. In this section, the multi-mode simulation approach is suggested to include the effects of space-charge neutralization for CW gyrotron operation. From the final beam parameters (after the desired neutralization, i.e. $V_{\mathrm{b}0}$, $V_{\mathrm{c}0}$, α_0 in Figure 4.5), the beam parameters before the neutralization (i.e. V_{b}, V_{c}, α) are calculated and based on it the two stage start-up simulation is performed. The step-by-step simulation approach is discussed below.

Step 1: select neutralization value

As a first step, the value of possible neutralization shall be selected, which indicates a reduction in the voltage depression due to the ionization of the residual gas particles. In Figure 4.5, the final beam voltage is equal to the cathode voltage. This scenario indicates the full neutralization (or 100 % neutralization) without any voltage depression. A variation of the beam parameters with 60 % neutralization is presented in Figure 4.6. Unlike the full-neutralization ($V_c = V_{b0}$), the cathode/acceleration voltage for the X percentage of neutralization can be calculated using $V_{\mathrm{c}} = V_{\mathrm{c}0} = V_{b0} + (1 - (X/100)) \cdot \Delta V_b$.

Step 2: calculate beam parameters after neutralization
Based on the cathode voltage (V_{c0}) calculated using step 1, the value of final relativistic factor (γ_0) and parallel normalized velocity ($\beta_\parallel$) after neutralization is calculated using following formulas,

$$\gamma_0 = 1 + \frac{V_{c0}}{511kV}, \tag{4.3}$$

$$\beta_{\parallel 0} = \sqrt{\frac{1 - \frac{1}{\gamma_0{}^2}}{1 + \alpha_0^2}}. \tag{4.4}$$

Here, α_0 is the desired velocity-ratio after the neutralization process.

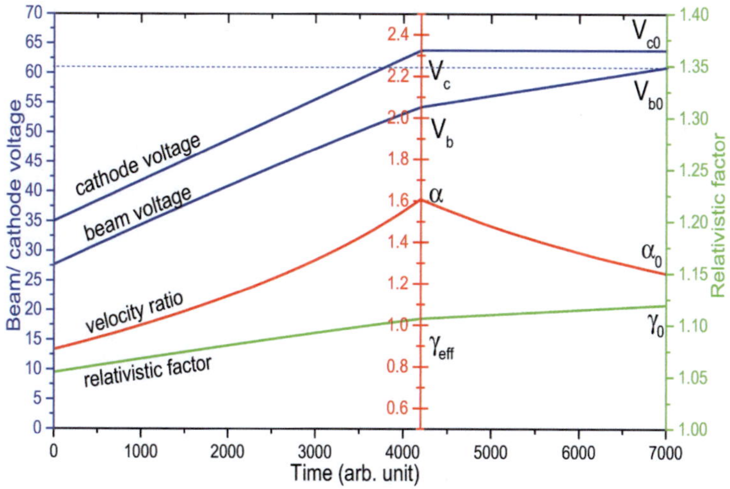

Figure 4.6: Using suggested technique, precise estimation of the beam parameters during neutralization process.

Step 3: estimate beam parameters before neutralization

The effective relativistic factor before neutralization factor can be calculated by solving following equation [BG91]

$$\gamma_{\text{eff}} = \gamma_0 - I^* \left[1 - \frac{\gamma_0^2 \left(1 - \beta_{\|0}^2\right)}{\gamma_{\text{eff}}^2} \right]^{-1/2}, \tag{4.5}$$

where,

$$I^* = 60 \frac{I_b}{511\ kV} \ln\left(\frac{r_0}{r_e}\right). \tag{4.6}$$

Considering the adiabatic approximation, perpendicular and parallel normalized velocity can be calculated using,

$$\beta_\perp = \frac{\gamma_0 \beta_{\perp 0}}{\gamma_{\text{eff}}} \tag{4.7}$$

$$\beta_\| = \sqrt{1 - 1/\gamma_{\text{eff}}^2 - \beta_\perp^2} \tag{4.8}$$

The velocity ratio before neutralization can be simply calculated by $\alpha = \beta_\perp/\beta_\|$. Because of the higher voltage depression, the value of γ_{eff} is lower than the γ_0, and eventually results in a higher value of the velocity ratio (α).

Step 4: initial start-up scenario

After calculating necessary parameters for the simulations, the multi-mode time-dependent start-up simulation is performed, in which cathode voltage is linearly increased from the noise level of the cathode voltage (V_{c0}) calculated using step 1. The pitch-factor is modified according to adiabatic equation. It should be noted that, in this step, the cathode voltage is controlled, instead of the beam energy. The beam energy at the end of step 4 is equal to the $V_b \sim (\gamma_{\text{eff}}-1) \cdot 511$.

Step 5: neutralization phase

In this second phase of simulation, the effects of space-charge neutralization are considered and beam parameters are changed accordingly. The beam energy and velocity ratio are varied from V_b and α to V_{b0} and α_0, respectively. Since the neutralization is a slow process, thus, for more realistic simulations, the beam parameters are modified in a relatively slow-time scales.

4.2.3 Simulation results for the DEMO gyrotron including neutralization

The suggested approach is verified for a 236 GHz DEMO gyrotron design and the final result is presented in Figure 4.7. In this case, 60 % neutralization is considered in analysis. The variation of the beam parameters during the neutralization process is presented in Figure 4.6. The calculated final cathode voltage (V_{c0}) and beam energy (V_{b0}) using step 1 are equal to 63.844 kV and 61 keV, respectively. Using the equations suggested in step 2 and 3, necessary beam parameters are calculated. The effective velocity ratio (α) and beam energy (V_b) before neutralization are 1.6 and 54.30 keV, respectively.

The first stage of simulation (step 4) is performed till t = 4200, in which cathode voltage is increased from 20 kV to 63.844 kV. In the second stage of the simulation (step 5), and due to neutralization, the beam energy is increased from 54.30 keV (t = 4200) to 61 keV (t = 7000) with velocity ratio variation from 1.60 to 1.25. Even including the effects of space-charge neutralization, the stable operation of $TE_{-43,15}$ is possible for a 236 GHz hollow-cavity design, which further validate the suggested physical design and operating parameters for CW operation.

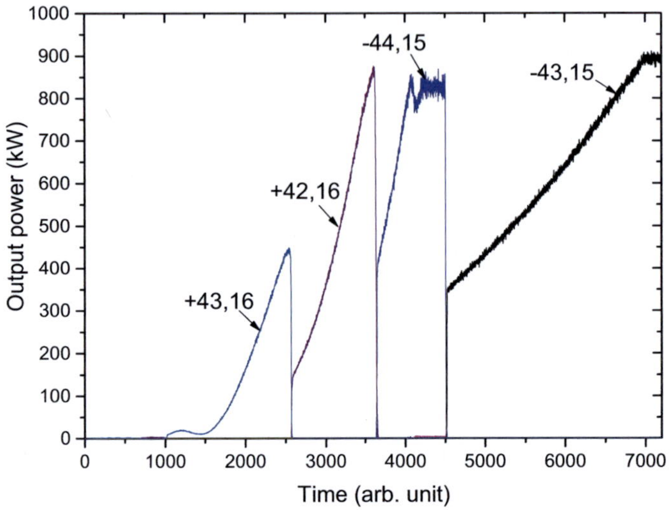

Figure 4.7: RF behavior of a 236 GHz, $TE_{-43,15}$ including space-charge neutralization effect.

5 Operational limits for a DEMO gyrotron

As discussed in previous chapters, using the selected operating mode - $TE_{-43,15}$ for a 236 GHz DEMO gyrotron, the estimated power from suggested hollow-cavity design is 920 kW with an interaction efficiency of 36 %. As per the EUROfusion baseline 2012, in total 50 MW power is required for plasma heating and current drive applications [GBB+15] [PTZ+12]. Less number of tubes per plant support reliable and cost-effective operation. Therefore, the high-output power per tube is desirable. In this chapter, by considering mode competition, the several generic approaches are suggested to identify the mode eigenvalue limit for stable operation. After investigating the eigenvalue limit, new cavity design and operating parameters are proposed for a 236 GHz, 1.5 MW gyrotron.

5.1 Importance of eigenvalue limits for high-power operation

The cavity wall-loading is highly dependent on the operating frequency ($P_{\Omega} \propto f^{2.5}$) (refer to equation 1.27) and the output power of high frequency gyrotrons ($f > 100$ GHz) is limited by the cavity wall-loading. From present capabilities of cavity cooling technologies, the maximum allowable cavity wall-loading is limited to 2 kW/cm^2. To improve the output power per tube, one possibility is to improve the cavity cooling system which eventually increases the maximum allowable wall-loading of the cavity. The effects of cavity wall-loading on the performance of a 236 GHz hollow-cavity gyrotron is presented in section 2.4.2. The high wall-loading capacity significantly increases the final output power. The investigation of the improved cavity cooling system is ongoing and is presented in [SBN$^+$15, SBN$^+$16].

The second possible solution for operation at high output power is to use higher order modes (e.g. eigenvalue > 103 for 236 GHz gyrotron). An eigenvalue is defined as the root of the derivative of the Bessel function for corresponding operating mode. Using a higher order mode, it is possible to increase the effective cavity area, which considerably reduces the maximum wall-loading. Of course, the mode spectrum density increases with high order modes and it might not be possible to excite the desired mode due to critically high mode competition. For a particular operating frequency, it is very important to know the eigenvalue limit for stable operation, which allows to choose the most suitable mode for high power operation.

For a 236 GHz DEMO gyrotron, the mode eigenvalue of the selected operating mode $TE_{-43,15}$ is 103.2132. The stability analysis is performed for the single-mode and the multi-mode cases, and the corresponding results are presented in Figure 5.1. The simulation set-up for stability analysis is similar to section 2.3.4. The operating parameters suggested in section 2.3.2 are used for this analysis (V_b = 58 keV, I_b = 39 A, B_0 = 9.130 T).

From the nominal operating point, the beam energy is increased step-wise till the operating mode losses stability. In both cases (single mode and multi-mode), the main mode is stable up to 60.7 keV which clearly indicate that, the mode loss is an effect of mode detuning, instead of mode competition.

As mode competition is under control for an eigenvalue of 103. It is needed to investigate the possibilities of high order mode with an eigenvalue more than 103 to further increase the output power per tube. The main aim of this analysis is to identify the eigenvalue limit for stable operation. Based on

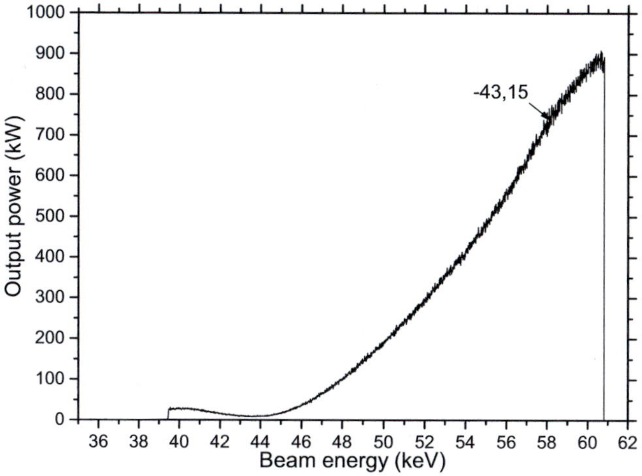

(a)

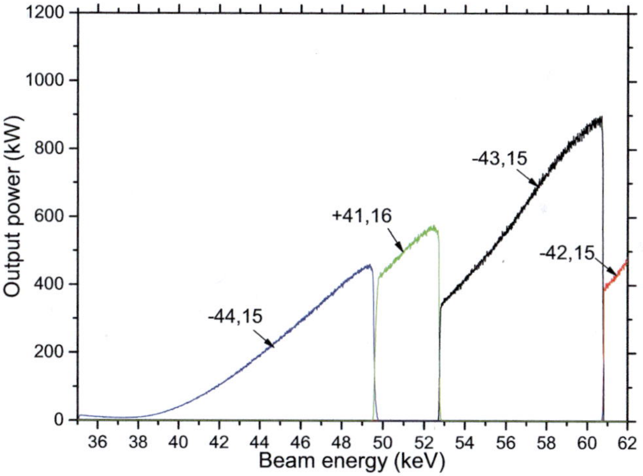

(b)

Figure 5.1: (a) Single-mode and (b) multi-mode, stability analysis with $TE_{-43,15}$ mode at 236 GHz. In both cases, the excitation of $TE_{-43,15}$ is possible till a beam energy of 61.7 keV.

this eigenvalue limit, the maximum possible output power can be estimated for a particular frequency. The methods to find eigenvalue limit is studied in this chapter. Those generic methods are also valid for all high-power future fusion gyrotrons.

5.2 Considerations for the analysis

The following conditions are considered for the eigenvalue limit analysis.

- Diode start-up scenario: The diode start-up scenario is selected in the analysis, in which the beam energy increases linearly during the start-up, while the velocity ratio varies according to the adiabatic approximation. The effect of advanced start-up scenario on gyrotron performance and eigenvalue limit will be discussed in the following chapter.
- Selection of neighboring modes: Within the frequency band of – 5 % to + 10 % (224.2 GHz to 259.6 GHz) of the center frequency - (236 GHz), all modes having relative coupling co-efficient more than 0.5 are considered. The coupling of the individual modes can be calculated using equation 1.21 and it is normalized to the main mode coupling to determine relative coupling co-efficient of individual mode.
- Multi-purpose operation of the gyrotron is not considered in this analysis.
- Physical parameters of the cavity: After the systematic hollow cavity design approach, the finalized cavity parameters suggested in Table 2.4 are used for this analysis. Of course, the cavity radius and beam radius are chosen according to the particular operating mode.

The time-dependent, self-consistent simulations and stability analyses are performed using Evridiki code of EURIDICE code-package [APIV12] and part of the simulations are performed using the computational resources of IFERC-CSC Helios supercomputer.

5.3 Selection of suitable modes

Before selecting the operating modes, it is very important to know the effects of mode properties on the gyrotron behavior. The selected modes for the hollow-cavity and co-axial cavity gyrotrons are plotted on the azimuthal to radial mode index plan (see Figure 5.2) and also listed in Table 5.1. The eigenvalue contours are also plotted with the difference of 50. Based on the relative mode caustic radius ($r_c = m/\chi_{m,n}$), the gyrotron modes are classified into three main types.

5.3.1 Type I: state-of-the-art hollow-cavity modes

During early gyrotron development, the main operating mode selection criteria is limited to the dimpled-wall Quasi-Optical Launcher (QOL) design [JTPR06]. The optimum performance of the dimpled-wall QOL with minimum stray radiation can be achieved only with the modes having relative caustic radius close to 0.5 [ADV08]. This type of modes for a 118 GHz, 140 GHz and 170 GHz gyrotrons are indicated with blue squares and the trend line is shown with a blue dotted line [PADT07] [SSH$^+$01] [Thu03] [CBB$^+$11]. For particular eigenvalue, when compared to the other two mode groups, these modes are having a low radial index. In other word, the mode maxima is relatively close to the cavity wall, which eventually leads to the high wall-loading. Due to the wall-loading considerations, the modes with relative caustic radius equal or higher than 0.5 are not suitable choice for high power operation at high frequency (> 200 GHz).

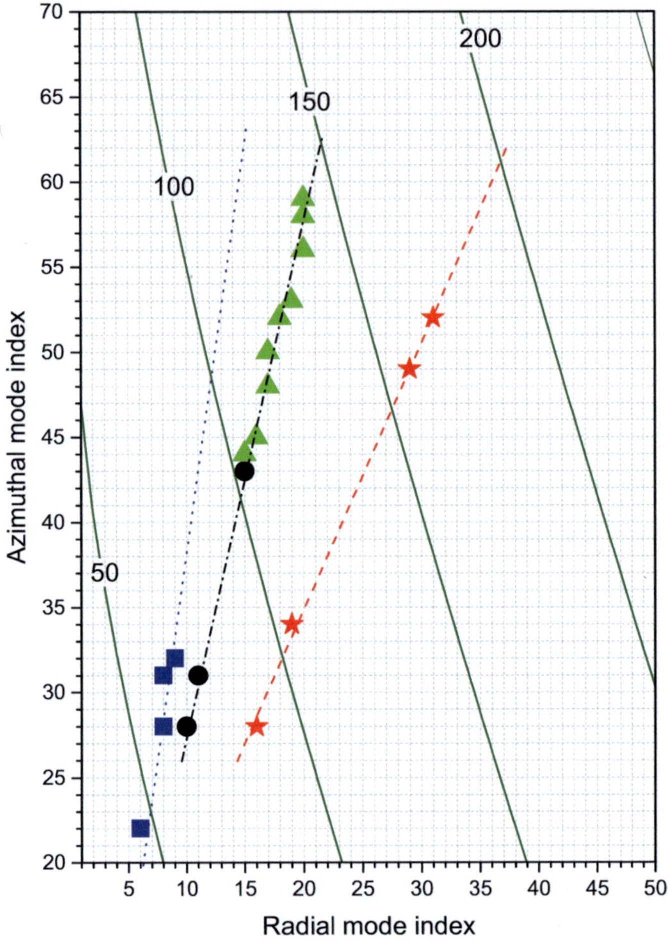

Figure 5.2: Modes of the state-of-the-art gyrotrons are plotted in plane of azimuthal index to radial index. Hollow cavity modes are shown with blue squares. Operating modes for the coaxial gyrotrons are indicated with red stars. The new type operating mode for recent gyrotrons are shown with black circles and the modes selected for the analysis are plotted with green triangles.

Table 5.1: Classification of gyrotron modes based on relative caustic radius.

operating mode	relative caustic radius
state-of-the-art hollow-cavity modes	
$TE_{-22,6}$	0.482
$TE_{-28,8}$	0.466
$TE_{-31,8}$	0.486
$TE_{-32,9}$	0.466
co-axial cavity modes	
$TE_{-28,16}$	0.320
$TE_{-34,19}$	0.322
$TE_{-49,29}$	0.310
$TE_{-52,31}$	0.310
new type of hollow-cavity modes	
$TE_{-28,10}$	0.417
$TE_{-31,11}$	0.417
$TE_{-43,15}$	0.416

5.3.2 Type II: coaxial-cavity modes

In a coaxial cavity gyrotron, using corrugated coaxial insert, the mode competition is controlled and the coupling with neighboring modes is suppressed, which allows excitation of the desired mode, even with dense mode spectrum. Due to the controlled mode competition, very high order operating mode can be used to increase output power per tube [PDD+04] [KHA+12]. In Figure 5.2, the selected modes for the coaxial cavity gyrotrons are presented with the red star and the trend line is also shown with a red dashed line. Compared to the hollow cavity modes, this type of modes are having high radial indexes with average relative caustic radius ($m/\chi_{m,n}$) is close to 0.3. Thus, these type of modes are more concentrated towards the cavity center, which significantly reduces the wall-loading. Due to small caustic radius, such modes are also desired to have a sufficient interaction with coaxial insert. Since the relative caustic radius is not close to 0.5, therefore the dimpled-wall quasi-optical launcher is not a suitable choice for efficient mode conversion. The linear output system [DIW$^+$04] or new type of launchers (mirror-line [JTP$^+$08] or hybrid-type [JTGJ16]) are used for efficient mode conversion.

In coaxial gyrotrons, the voltage depression is reduced by using the grounded insert. However, for the hollow-cavity gyrotrons, due to large differences between cavity radius and beam radius, the voltage depression is critically high (according to Equation 4.1). For long-pulse gyrotron operation (as discussed in section 4.2), the beam parameters are changed critically because of high depression voltage and lead to unstable mode excitation. So, considering the limitations of high voltage depression, this type of mode is not selected for the limit analysis of a 236 GHz hollow-cavity DEMO gyrotron.

5.3.3 Type III: new family of hollow-cavity modes

The development of new advanced launchers provides the flexibility to choose the operating mode having relative caustic radius other than 0.5 [JTGJ16]. The modes having relative caustic radius close to 0.4 are selected and used successfully in various experiments [IOK$^+$16]. These type of modes are presented in Figure 5.2 with black circles.

Table 5.2: Selected modes and respective operating parameters for the eigenvalue limit analysis.

Case name	DM1	DM2	DM3	DM4	DM5	DM6	DM7	DM8	DM9
Eigenvalue	104.46	109.17	116.4	118.91	124.87	129.58	136.8	139.32	144.02
Cavity Mode	$TE_{-44,15}$	$TE_{-46,16}$	$TE_{-48,17}$	$TE_{-50,17}$	$TE_{-52,18}$	$TE_{-53,19}$	$TE_{-56,20}$	$TE_{-58,20}$	$TE_{-59,21}$
Cavity radius (mm)	21.14	22.09	23.55	24.06	25.26	26.22	27.68	28.19	29.14
Beam radius (mm)	9.28	9.49	10.1	10.51	10.93	11.13	11.75	12.16	12.37
Rel. Caustic radius	0.4211	0.4121	0.41237	0.42048	0.41641	0.409	0.40933	0.4163	0.40964
operating parametrs									
Q	1445	1457	1483	1493	1573	1533	1553	1559	1567
Beam energy (keV)	64	67	68	69	71	73	76	78	80
Magnetic field (T)	9.198	9.244	9.259	9.275	9.303	9.335	9.373	9.403	9.435
Beam current (A)	43	45	50	51	55	58	62	64	66
Output power (kW)	1050	1140	1294	1325	1470	1547	1710	1805	1900
Efficiency (%)	38	38	38	38	38	38	37	37	36
Wall loading (kW/cm^2)	2	1.96	1.98	1.97	1.98	1.97	1.97	2	2

The effective wall-loading of these modes are comparatively less than the type I modes (relative caustic radius ~ 0.5). In the hollow-cavity case, the voltage depression such modes are also less than the type II modes, which are having relative caustic radius of 0.3. Due to the moderate wall-loading and voltage depression, this type of modes are preferred for hollow-cavity design. These features also justify the selection of $TE_{-43,15}$ mode for a 236 GHz DEMO gyrotron.

For the eigenvalue limit analysis, type III modes are selected due to discussed advantages. The list of the selected modes is presented in Table 5.2 and plotted in Figure. 5.2 with green triangles. Within an eigenvalue range of 105 to 145, nine modes are selected with approximate eigenvalue separation of 5 and having a relative caustic radius of 0.41. The selected modes are named as 'DM1' to 'DM9' in ascending order of eigenvalue. The different cavity geometries are considered in the analysis, in which the radius of interaction section is selected according to the operating modes. All other physical parameters of the cavity are identical for the selected cases (L_1 / L_2 / L_3 = 16 mm / 12 mm / 16 mm, θ_1 / θ_3 = 2.5 ° / 2.0 °, D_1 / D_2 = 4 mm / 5 mm). With the help of single mode time-dependent simulations, the operating parameters (magnetic field, beam energy and beam current) are finalized and are enlisted in the Table 5.2. The velocity ratio (α) of 1.25 and cavity wall-loading of 2 kW/cm^2 is considered for all modes. The operating parameters are finalized to achieve minimum interaction efficiency of 36 %.

The cavity radius (r_{cav}) increases with the eigenvalue, which increase the total power handling capacity of the cavity for the same wall-loading. Due to same relative caustic radius, the behavior of these modes is identical, except the mode competition. A mode competition increases with the increase in eigenvalue. The mode spectrum of two modes DM1 ($\chi_{m,n}$ = 104) and DM9 ($\chi_{m,n}$ = 145) are compared in Figure 5.3. Within the selected bandwidth of -5 % to +10 %, the modes having relative coupling more than 0.5 are plot-

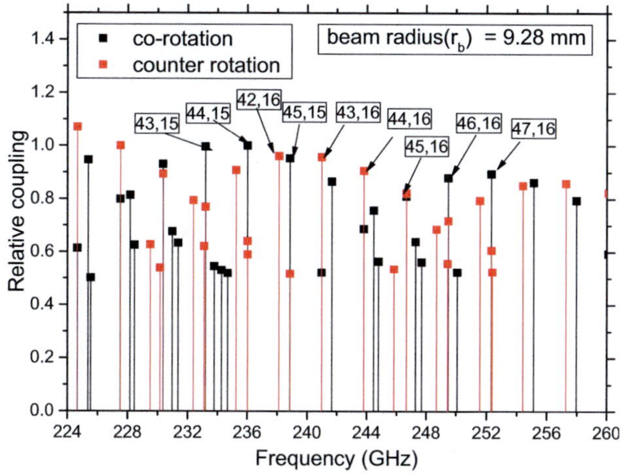

(a)

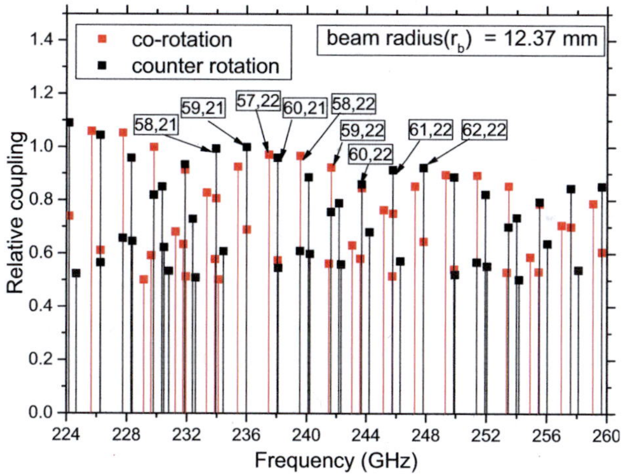

(b)

Figure 5.3: Comparison of mode spectrum of the (a) DM1 ($\chi_{m,n} \sim 105$) and (b) DM9 ($\chi_{m,n} \sim 145$) case.

ted. The number of modes within selected bandwidth increase with mode eigenvalue. The main aim of this analysis is to identify the eigenvalue limit for stable operation. Based on this eigenvalue limit, the maximum possible output power can be estimated for a particular frequency.

5.4 Effect of detuning on mode stability

To determine eigenvalue limit, the RF behavior and stability of a 236 GHz gyrotron is studied with the selected modes in Table 5.2. For the gyrotron operation with high eigenvalue mode, the effect of detuning and critical mode competition are the two main reasons, which suppress the excitation of the operating mode. In section 5.5 and 5.6, the effects of mode competition and spectral density on mode stability are analyzed with non-linear and linear theories.

The three modes from Table 5.2 are selected to study the detuning characteristics at high eigenvalue and its effects on mode stability. Since the desired window of the eigenvalue limit analysis is in between 105 and 145. The first selected mode is $TE_{-44,15}$ (DM1) with an eigenvalue of around 105, the second mode is $TE_{-52,18}$ (DM5) with an eigenvalue of around 125 and third selected mode is $TE_{-59,21}$ (DM9) with an eigenvalue close to 145.

The influence of frequency slippage (detuning) on eigenvalue limit is presented in this section. The normalized detuning can be calculated using,

$$\Delta = \frac{2}{\beta_\perp^2}\left(\frac{\omega - s\cdot\Omega_c}{\omega}\right) = \frac{2}{\beta_\perp^2}\left(\frac{\omega - s\cdot\left(\frac{e\cdot B_0}{\gamma_0\cdot m_e}\right)}{\omega}\right). \tag{5.1}$$

For the optimum gyrotron operation, the typical value of detuning is between 0.4 and 0.5 [DT86]. In this study, using a magnetic field sweeping,

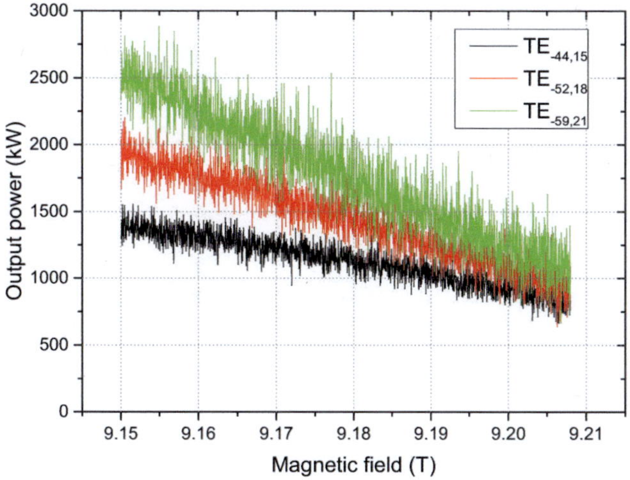

Figure 5.4: Detuning analysis with the modes: DM1 ($TE_{-44,15}$), DM5 ($TE_{-52,18}$) and DM9 ($TE_{-59,21}$).

Table 5.3: Operating parameters for detuning analysis.

Mode	Magnetic field (T)	Beam energy (keV)	Beam current (A)	Wall-loading (kW/cm^2
$TE_{-44,15}$ (DM1)	9.028	61	58	2
$TE_{-52,18}$ (DM5)	9.028	61	81	2
$TE_{-59,21}$ (DM9)	9.028	61	110	2

the normalized detuning factor is changed from 0.4 to 0.5 and the performance of the gyrotron is estimated. For the direct comparison of the detuning effect, the beam energy is kept constant for all three modes and beam current is selected considering the maximum allowable wall-loading. Thus, the value of detuning is directly modulated by the magnetic field. The selected operating parameters for the detuning analysis is listed in Table 5.3.

The single mode self-consistent simulations are performed to study effect of detuning on mode stability. The magnetic field is swept from 9.208 T to 9.150 T, which corresponds to the sweep in normalized detuning from 0.4 to 0.5. As a first step, the individual mode is excited considering only a single mode at magnetic field of 9.208 T (normalized detuning $\sim$ 0.4). Then the magnetic field is reduced to 9.150 T. The result of this analysis is presented in Figure 5.4. The excitation of all three modes (DM1, DM5 and DM9) is stable over the detuning sweep. From the result, it is clear that, within the selected eigenvalue range (105 to 145) and the operating range of normalized detuning (0.4 to 0.5), the behavior of all selected modes are identical and the only possible reason for the mode instability is critical mode competition. After having shown that detuning is irrelevant, the next two sections are dedicated to study the effects of mode competition on eigenvalue limit.

5.5 Eigenvalue limit analysis using multi-mode simulations

Using nonlinear theory of gyrotron interaction [DT86] [Nus04] [SANL93] [Bor91], two generalized approaches for the eigenvalue limit estimations are suggested and discussed in this section. The operating modes listed in Table 5.2 are selected for this analysis.

5.5.1 Time-dependent multi-mode self-consistent analysis

Stability of the operating mode is investigated self-consistently with respect to the effects of neighboring modes. This type of calculation estimates the power transfer from the electron beam to the individual modes, keeping a self-consistent equilibrium. For stable gyrotron operation, at steady-state condition, the mode amplitude of the operating mode should be significantly higher, while keeping the amplitude of neighboring modes less than the computational noise level. If the operating modes are outside the eigenvalue limit, it is not possible to obtain the stable mode excitation due to high mode competition and close neighboring mode excites.

Initially, the optimized operating parameters for the individual mode are finalized considering the only single mode scenario without including the effect of neighboring modes in self-consistent simulations. The realistic electron beam parameters have been used with the Gaussian velocity spread of 6 % (rms) and the total radial width of $\lambda/4$. The magnetic field is fixed to have stable operation with the stability margin of 2 keV. In the single mode case, the stable operation is achieved with all selected modes (DM1 - DM9), which also support detuning analysis presented in section 5.4. After finalizing the operating parameters from single mode simulations, the multi-mode time-dependent simulations are performed taking into account all relevant neighboring modes. All the modes within 224 GHz and 260 GHz frequency band (-5 % to +10 % of 236 GHz) and having coupling co-efficient of more than 0.5 are selected for the time-dependent simulations.

For the mode with lowest eigenvalue of 105 (DM1), the number of selected neighboring modes are 55 (see Figure 5.3 (a)), which is increased to 85 in the case of operating mode with the highest eigenvalue of 145 (DM9) (Figure 5.3 (b)). As compared to the single mode simulations, the computational

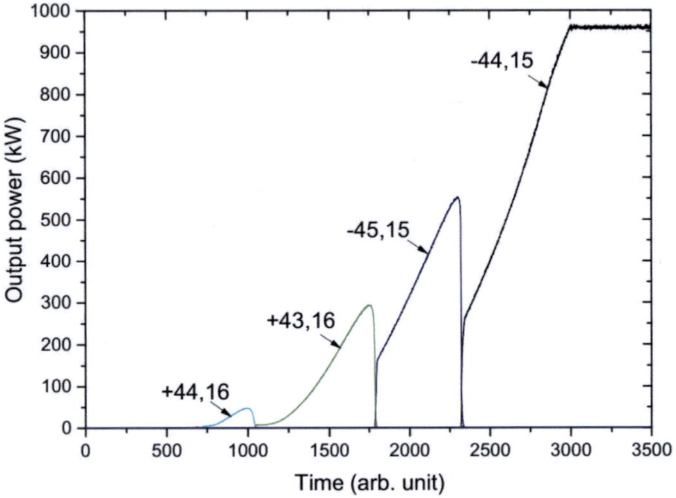

Figure 5.5: Multi-mode analysis with DM1 ($TE_{-44,15}$) mode.

time of such self-consistent simulations with large numbers of neighboring modes are significantly increased. For example, the computational time for particular simulation is more than 15 hours with 320 processors. Of course, the computational time is strongly dependent on number of particle used for the simulations.

In Figure 5.5, the result of multi-mode simulation is presented with the DM1 (eigenvalue $\sim$ 105) mode and 55 neighboring modes. The beam energy changes from 20 to 64 with the adiabatic change of beam velocity ratio and beam current. The rest of the operating parameters for the simulations of DM1 mode are enlisted in the Table 5.2. The presented result supports the stable excitation of $TE_{-44,15}$ mode without any parasitic mode excitation. The mode competition is under control and the main mode can be excited using diode-type start-up scenario.

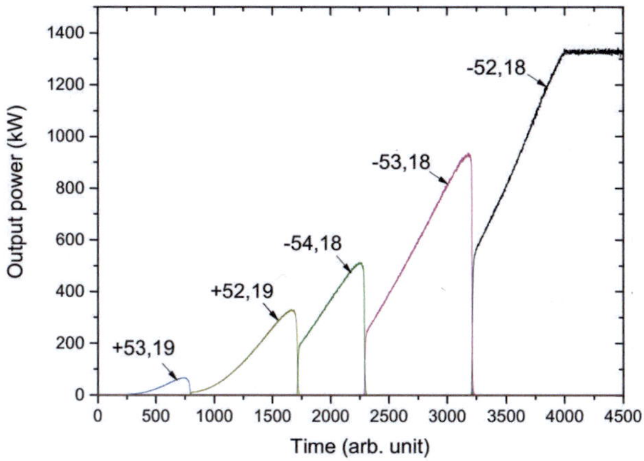

(a)

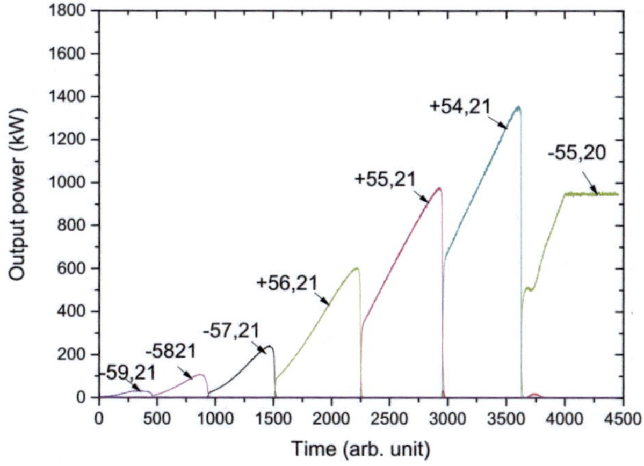

(b)

Figure 5.6: Eigenvalue limit investigation using non-linear self-consistent multi-mode simulations. (a) Stable mode excitation is possible with DM5 (eigenvalue $\sim$ 125), while due to high mode competition the desired mode $TE_{-56,20}$ is not excited in (b).

After investigating the performance of the DM1 mode with the eigenvalue of 105 (lowest from the selected modes), as a next step, the behavior of DM9 (highest eigenvalue from the selected modes) is verified. Due to the critical mode competition, it is not possible to excite desired operating mode ($TE_{-59,21}$) during the start-up and the power of the next neighboring mode ($TE_{-58,21}$) is raised (see Figure 6.3 (a)). It is clear from the results that, for diode start-up scenario the mode DM9 (eigenvalue $\sim$ 145) is beyond the eigenvalue limit of 236 GHz gyrotron and eigenvalue limit lay in between the mode eigenvalue 105 and 145.

In Figure 5.6, the RF behavior of the DM5 mode (eigenvalue $\sim$ 125) and DM7 (eigenvalue $\sim$ 135) are compared. For DM5 case, the beam energy is increased linearly from 30 keV (t = 0) to 71 keV (t = 4000) and steady-state condition is selected from the time-interval t = 4000 to t = 4500. The variation of beam energy is from 33 keV to 76 keV in DM7 case (Figure 5.6 (b)). For the simulations of DM5 mode, 70 neighboring modes are included, while in DM7, in total 76 neighboring modes are included. In the case of DM5, the stable excitation of the operating mode $TE_{-52,18}$ is possible with controlled mode competition. On the other hand, because of the dense mode spectra, the desired operating mode ($TE_{-56,20}$) is not excited in the case of DM7.

The similar multi-mode time-dependent simulations are also performed with the modes having eigenvalue between 105 and 125 (DM2, DM3 and DM4) and in each case, the desired operating mode excites successfully in multi-mode start-up simulations. On other hand, with the operating mode having eigenvalue more than 125 (DM6 – DM9), it is not possible to excite desired operating mode with the diode start-up scenario. So, this analysis strongly suggests the eigenvalue limit of 125 for the 236 GHz hollow-cavity DEMO gyrotron. With the help of time-dependent multi-mode, self-consistent simulations, the effect of parasitic modes can be studied and it is possible to estimate the eigenvalue limit for high power gyrotron operation.

For the multi-mode time-dependent simulations, the selection of the neighboring mode is critical for the precise estimation of the eigenvalue limit. In Table 5.4, the effect of the number of neighboring modes selected for the multi-mode analysis is presented for different cases. With consideration of only few neighboring modes (20 modes), the simulation results indicate the excitation of the operating mode till the eigenvalue of 140 (DM8). When all possible parasitic modes having a relative beam coupling more than 0.5 are used in the simulations, then it is not possible to excite desired operating mode with the eigenvalue more than 125 (DM5). So, for the accurate estimation of the eigenvalue limit, it is necessary to include a sufficiently large number of neighboring modes in the multi-mode time-dependent simulations.

Table 5.4: Effect of the number of neighboring modes selected for the multi-mode analysis for the DM7, DM8 and DM9 case. All possible neighboring modes include the modes having relative coupling of more than 0.5. The number of such modes are 76, 79 and 85 for DM7, DM8 and DM9, respectively.

Multi-mode simulations	Single mode	< 20 neighboring modes	All possible neighboring modes
DM7	stable excitation	stable excitation	mode loss
DM8	stable excitation	stable excitation	mode loss
DM9	stable excitation	mode loss	mode loss

5.5.2 Stability analysis

In this section, the effect of mode competition is determined by comparing the stability margin of single mode operation with multi-mode operation. Based on that, the mode eigenvalue of a 236 GHz DEMO gyrotron is determined. The whole process is sub-divided into the following three steps.

Step I: As single mode excitation is not affected by the mode competition, for each case, the single mode start-up simulation is performed using the operating parameters presented in Table 5.2. Due to single mode considerations, it is possible to excite all the selected mode (DM1 – DM9) with optimum output power and efficiency. These intermediate results are further used in the step II and III of this method.

Step II: The main aim of this analysis is to identify the bean energy at which the operating mode loses the stability due to high detuning in the single mode case. From the final stage of step I, the single mode self-consistent simulation is continued with the step-wise increase in beam energy.

Step III: In this step, after single mode simulations of the step I, all the possible neighboring modes are considered in the simulation and beam energy is increased step-wise to identify the mode loss in the presence of neighboring modes.

When the mode loss in step II is at the similar beam energy as in step III, the operating mode is marked as a mode within the eigenvalue limit for diode start-up. In such case, the effects of mode competition is not significant and the mode lose is due to high detuning. When mode loss is at different beam energy in single mode (step II) and multi-mode (step III) scenario, it is clear that, the stability of the operating mode is influenced by the mode competition and the mode is outside the eigenvalue limit.

For the operating mode DM5 (eigenvalue $\sim$ 125) and DM7 (eigenvalue $\sim$ 135), the results of stability analysis are compared in Figure 5.7. In case of DM5, the results of stage II and stage III are presented in Figure 5.7 (a). From the nominal beam energy of 71 keV, the beam energy is increased in steps of 0.1 keV. The mode loss in the single mode case and multi-mode case are at the same beam energy. This indicates that, the mode loss in the case of DM5 is mainly because of the detuning. While in the case of DM7 (see Figure 5.7 (b)), the mode loss for the multi-mode scenario is prior to the single-mode scenario, which clearly indicate the effects of critical mode competition. From this analysis, it is clear that the mode DM5 (eigenvalue $\sim$ 125) is inside the eigenvalue limit and the mode DM7 (eigenvalue $\sim$ 135) is outside the eigenvalue limit. Therefore, the mode DM7 is not a suitable choice for the diode type start-up.

This method is also verified with DM1, DM6 and DM9 modes and the results are enlisted in Table 5.5. For the modes with eigenvalue of more than 125, the mode loss in the multi-mode case (step III) is prior to the mode loss in single-mode considerations (step II). The results validate the eigenvalue limit of 125 for a 236 GHz DEMO gyrotron and also support the findings of the limit of multi-mode analysis (section 5.5.1).

5.6 Eigenvalue limit using linear gyrotron theories

In this section, based on the linear theory of gyrotron interaction two methods are investigated for the mode eigenvalue estimation. These theories are based on mode-beam coupling and the effect of variation of the axial-field

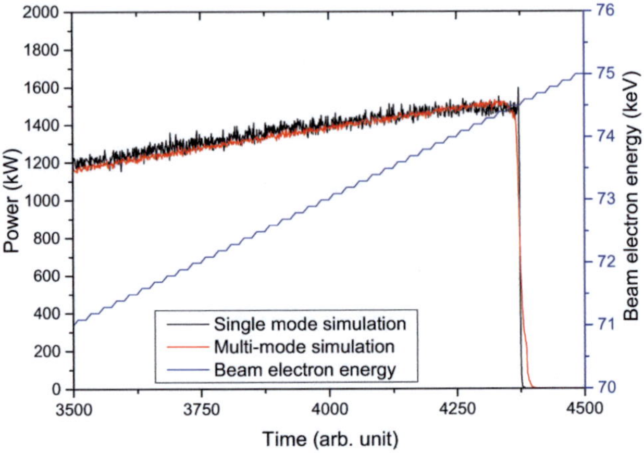

(a)

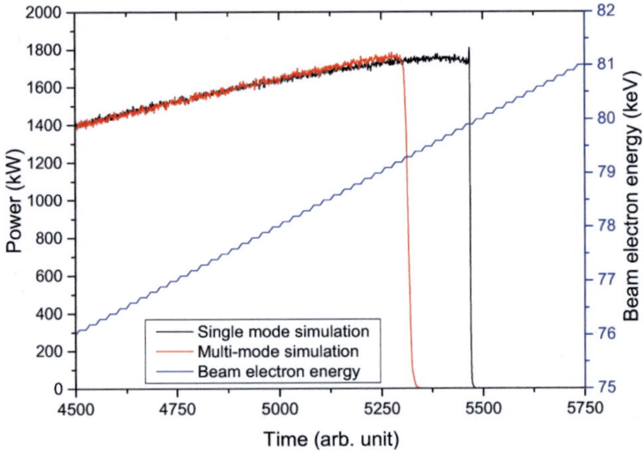

(b)

Figure 5.7: Stability analysis with the mode (a) DM5 (eigenvalue $\sim$ 125) and (b) DM7 (eigenvalue $\sim$ 135).

Table 5.5: Results of the stability analysis. For an eigenvalue higher than 125, mode loss in the multi-mode scenario is prior to the single mode case, which clearly indicates the influence of excessive mode competition.

Operating mode	Nominal beam voltage (keV)	Single mode		Multi mode	
		Mode loss (keV)	Stability margin	Mode loss (keV)	Stability margin
DM1	64	67.1	3.1	67.1	3.1
DM5	71	74.5	3.5	74.5	3.5
DM6	73	77.2	4.2	76.6	3.6
DM7	76	79.9	3.9	79.3	3.3
DM9	80	83.6	3.6	82.9	2.9

profile are not included. However, these methods are quick and suitable for the first order estimation.

5.6.1 Starting current plots

For specific operating parameters, the starting current represents the minimum beam current required to excite particular mode. It can be represented as a function of magnetic field or beam energy. In the case of soft-excitation scenario, the modes having minimum value of the starting current excite at steady-state. In addition to that, the starting current curves are also helpful to estimate possible modes during start-up. In this analysis, the starting current plots are used to estimate eigenvalue limit of high power gyrotrons and also to validate findings of eigenvalue limit based on non-linear theory as presented in section 5.5.

The starting current plots can be calculated using [Bor91] [BJ88],

$$
\begin{aligned}
1/I_{st} &= -QZ_0\left(\pi/\lambda \int \left|\widehat{f}(z)\right|^2 dz\right)^{-1} (e/8\gamma_0 m_e c^2) \\
&\cdot \; (k_{\perp \mathrm{m,n}} C_{\mathrm{m,n}} G_{\mathrm{m,n}} / \beta_{\|0})^2 \\
&\cdot \; [1+\omega_{\mathrm{RF}}\beta_{\perp 0}/2v_{z0}\cdot \partial/\partial\Delta_1] \left| \int_0^L \left(\left|\widehat{f}(z)\right|^2 e^{i\Delta_1 z} dz \right) \right|,
\end{aligned}
\tag{5.2}
$$

where, the effect of detuning is included using term,

$$
\Delta_1(z) = \frac{\omega}{v_{\|0}} \left(1 - \frac{\Omega_0(z)}{\omega_{RF}\gamma_0}\right), \tag{5.3}
$$

and the beam-field coupling coefficient can be defined as,

$$
\pi(C_{\mathrm{m,n}} G_{\mathrm{m,n}}) = \frac{J^2_{\mathrm{m}\mp 1}(k_{\perp \mathrm{m,n}} r_{\mathrm{b}})}{J^2_{\mathrm{m}}(\chi_{\mathrm{m,n}})(\chi^2_{\mathrm{m,n}} - m^2)}, \tag{5.4}
$$

where Z_0 is the free space impedance. The axial field profiles ($\widehat{f}$) is used to determined the starting current plot of particular mode. The detailed description of the equation 5.2 is discussed in [Bor91] [KT80] [BJ88].

The behavior of starting current plots for the selected higher order modes (Table 5.2) is studied with great focus on the effects of mode competition. The starting-current curves for $TE_{-52,18}$ (DM5 case, eigenvalue = 125) and $TE_{-56,20}$ (DM7 case, eigenvalue = 135) modes are compared in Figure 5.8. The starting current curves of the main operating modes are presented along with the eight parasitic modes. The selected neighboring modes are close to the main mode and have high beam coupling. The co-rotating modes are

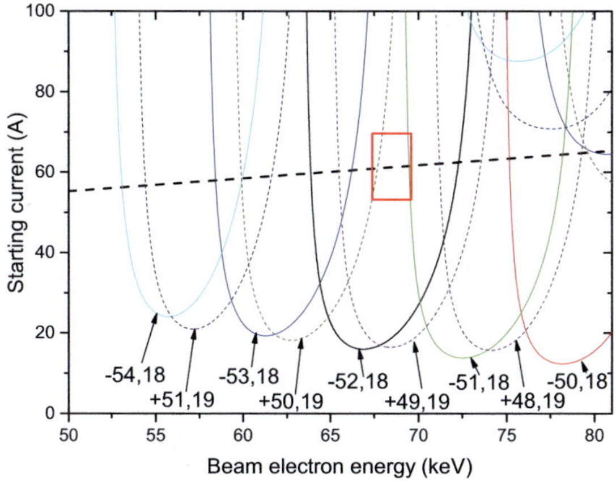

(a)

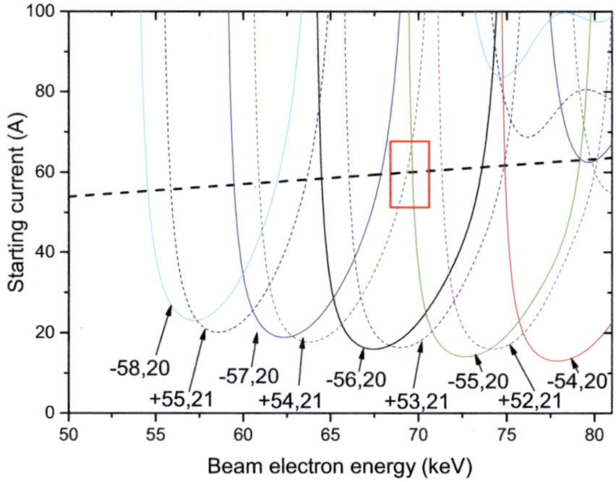

(b)

Figure 5.8: Starting current plot for (a) DM5 and (b) DM7.

presented with the solid lines, the counter rotating modes are plotted with the dashed lines and the beam line is presented with the black dashed line.

During the start-up scenario for a particular beam voltage, the modes with a minimum starting current are excited and remains excited till the starting current is lower than the beam current. As the beam energy increases, due to detuning effect, the value of starting current of the excited mode increases and losses its stability. For an easy understanding of mode excitation, the multi-mode start-up scenario of DM5 (Figure 5.6 (a)) and starting current plots (Figure 5.8 (a)) shall be compared. For example, with the beam energy of 55 keV, $TE_{-54,18}$ starts to excite and remains excited till the beam energy of around 60 keV. As the starting current of $TE_{-53,18}$ is minimum at beam energy of 60 keV, the mode $TE_{-53,18}$ is excited after $TE_{-54,18}$.

In case of DM5, the higher-frequency neighboring modes, $TE_{-(m+1),n}$ and $TE_{+(m-2),(n+1)}$ are well separated from the lower-frequency neighbor - $TE_{-(m-1),n}$, supporting excitation of the main mode in between (see red box in Figure 5.8(a)). For DM7, starting current curves of higher-frequency mode $TE_{+(m-2),(n+1)}$ and lower-frequency mode $TE_{-(m-1),n}$ are intersecting (see red box in 5.8(b)), i.e. one is excited just after the other breaks down, leading to a possible skip over the main mode during startup.

For a 236 GHz DEMO gyrotron, the finding of starting current analysis also supports the estimated eigenvalue limit of 125 from the non-linear theories. Till the mode eigenvalue of 125, starting current plots of next neighboring modes are well separated, which supports the stable main mode operation. With the diode-type start-up scenario, the mode competition is critical above mode eigenvalue of 125 and it is difficult to excite desired operating mode.

5.6.2 Mode spectrum analysis

In this part, the effect of mode competition is studied by analyzing the mode spectrum of the individual case. The coupling of electron beam with the individual cavity mode can be estimated using equation 1.21. The coupling of each mode is normalized to the main mode coupling and the relative coupling is used for an easy comparison. Figure 5.9 represents the mode spectra for the various modes. The modes within the bandwidth of –5 % to +10 % of the operating frequency of 236 GHz and having a relative coupling of co-efficient more than 0.5 are selected for analysis. A relative coupling of modes are also dependent on the selected beam radius. The beam radius for the maximum coupling is used for the individual case (see equation 1.22).

As the mode eigenvalue increases from 105 to 145, the number of modes having mode coupling more than 0.5 (within the selected bandwidth) increases (e.g. DM1 ($\chi_{mn} \sim 105$) = 55, DM5 ($\chi_{mn} \sim 125$) = 70, DM7 ($\chi_{mn} \sim 135$) = 76, DM9 ($\chi_{mn} \sim 145$)= 85), which clearly indicates the increase in mode competition with the mode eigenvalue. In Figure 5.9, the mode spectra of DM5 and DM7 are compared. The co-rotating modes are plotted with black color and the counter-rotating modes are plotted in red color. After detailed analysis of the mode spectra, the particular patterns were found in the variation of mode coupling, as described below.

For example, in case of DM5 (Figure 5.9 (a)), the co- and counter rotating modes having equal radial index are linked with bold solid lines. The co-rotating modes $TE_{XX,19}$ are linked with black line and counter-rotating $TE_{XX,19}$ modes are linked with red line. Both lines intersect at the mode $TE_{-53,19}$, where the coupling of co-rotating mode and counter-rotating mode is same (frequency $\sim$ 245 GHz). For high order modes, there is an individual cluster of modes for particular radial index (for example, in Figure 5.9 (b), the mode cluster of $TE_{XX,21}$ is presented). For all the selected cases in the eigenvalue estimation analysis, the pattern of mode spectra is similar and in each case, mode $TE_{m+1,n+1}$ is having equal co- and counter rotating mode coupling. In Table 5.6, the positions of mode having equal coupling are enlisted for different cases. It is clear that, the position of the modes with equal co- and counter rotating coupling move closer to the operating frequency with increasing eigenvalue.

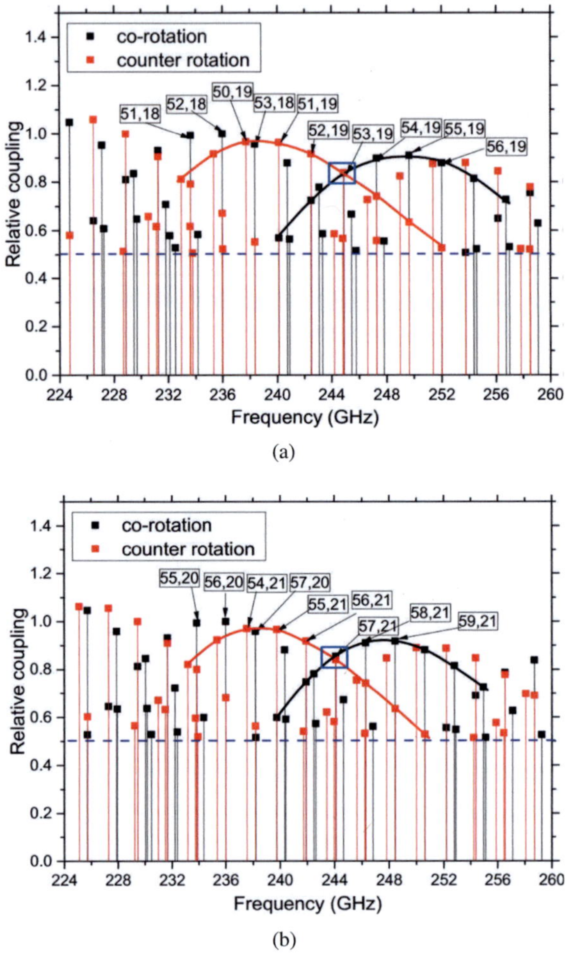

Figure 5.9: Mode spectrum of (a) DM5 and (b) DM7 case. The cluster of modes having same radial index are linked with solid lines.

Table 5.6: Influence of eigenvalue on the mode spectrum.

Case (eigenvalue)	Status	Position of mode with equal co-rotation and counter rotation mode
DM1 (104.46)	stable output	246.5
DM5 (124.87)	stable output	245
DM7 (136.80)	non-stable output	244
DM9 (144.02)	non stable output	243

The following observations are made by comparing the results of multi-mode start-up scenario (Figure 5.6) with the corresponding mode spectra (Figure 5.9). When the excited modes during the start-up have higher frequency than the frequency of equal coupling of co- and counter rotating mode ($TE_{m+1,n+1}$), it is not possible to excite the desired operating mode. For example, in case of DM7 (Figure 5.6 (b)), initial excitation of $TE_{-59,21}$ leads to the excitation of counter-rotating $TE_{+57,21}$ mode followed by the mode series ($TE_{+56,21}$, $TE_{+55,21}$, $TE_{+54,21}$) and suppresses the desired operating mode $TE_{-56,20}$. Similarly, in case of DM6 to DM9 (eigenvalue > 125), the modes with frequency higher than $TE_{m+1,n+1}$ have been generated during the start-up, and for each case, $TE_{m-1,n}$ mode excites at a steady state instead of desired $TE_{m,n}$ mode. In summery, it is not possible to determine the eigenvalue limit only by using mode spectra. However, the mode spectra analysis is very useful to study the trend of mode excitation during start-up.

Based on the linear and non-linear theories of gyrotron interaction, different approaches are suggested to find the mode eigenvalue limit for stable gyrotron operation. Considering the eigenvalue limit, the suitable operating mode shall be selected for maximum possible output power. From all the generic methods suggested in section 5.5 and 5.6, indicates the same eigenvalue limit of 125 for a 236 GHz gyrotron for simple diode-type start-up scenario. Below mode eigenvalue 125, the stable mode excitation is possible with the controlled mode competition. Along with suggested modes in Table 5.2, stability of many intermediate modes have also been tested, which also support the estimated eigenvalue limit.

5.7 236 GHz, 1.5 MW gyrotron design

As suggested in previous sections, with diode-type start-up, the stable mode excitation is possible till mode eigenvalue of 125. In this section, the physical parameters of the cavity and the operating parameters of gyrotron are optimized to get 1.5 MW output power with operating mode $TE_{-52,18}$ (eigenvalue $\sim$ 125). As compared to our standard design with $TE_{-43,15}$ mode, it is possible to get more than 1.5 times of the output power with the same wall loading of 2 kW/cm^2.

For the 1.5 MW design, new physical parameters of the cavity are enlisted in Table 5.7. Most of the physical parameters are same as the standard design except the cavity interaction length (L_2). Here, the value of the interaction length is selected to 11.5 mm ($\sim 9 \cdot \lambda$), instead of 12 mm as in the standard design. As discussed in section 2.2, the reduced interaction length further improves the mode stability, with increased output power along with slight reduction in the interaction efficiency [Nus04].

Considering realistic beam parameters (perpendicular velocity spread = 6 % (rms) and radial width of $\lambda/4$), the multi-mode start-up scenario with 70 parasitic modes are presented in Figure 5.10. The finalized operating parameters are also listed in Table 5.7. The beam energy increases linearly from 39 keV (t = 0) to 74 keV (t = 3000). The stable $TE_{-52,18}$ mode is excited at steady-state (t = 3000 to 3500) without any excitation of spurious modes. With the modified cavity profile and the optimized operating parameters, the output power is increased from 1320 kW (Figure 5.6) to 1530 kW (Figure 5.10). Using suggested approach in section 4.2, performance of the cavity is also verified including the effects of space-charge neutralization. The stable gyrotron operation has been achieved with 60 % of neutralization.

Table 5.7: Physical parameters of the cavity and gyrotron operating parameters of the new 236 GHz, 1.5 MW design are compared with original 1 MW design.

Parameters	236 GHz, 1 MW design (original design)	236 GHz, 1.5 MW design (new design)
Interaction mode	$TE_{-43,15}$	$TE_{-52,18}$
Eigenvalue	103.21	124.87
Physical parameters of cavity		
Cavity radius (mm)	20.88	25.26
L_1 / L_2 / L_3 (mm)	16 / 12 / 16	16 / 11.5 / 16
D_1 / D_2 (mm)	2 / 2.5	2 / 2.5
θ_1 / θ_2 / θ_3	2.5 ° / 0 ° / 2 °	2.5 ° / 0 ° / 2 °
Operating parameters		
Beam radius (mm)	9.06	10.93
Beam current (A)	43	63
Beam energy (keV)	61	74
Magnetic field (T)	9.165	9.355

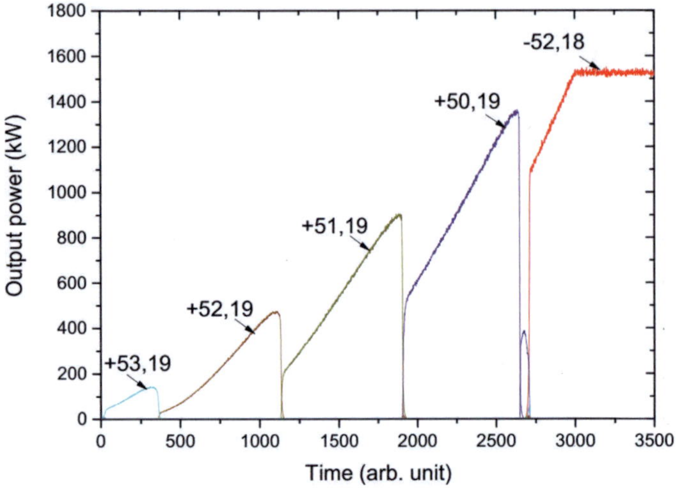

Figure 5.10: Multi-mode start-up scenario for new 236 GHz, 1.5 MW design. The stable operating of $TE_{-52,18}$ mode is possible with optimized cavity design and suggested operating parameters.

Tolerance to beam misalignment

Using the macro-electron beam method proposed in section 4.1.3, the electron beam misalignment tolerance analysis has been performed for the new 1.5 MW gyrotron design.

Due to increase in the mode density at eigenvalue of 125, the beam misalignment tolerance for the stable mode excitation decreases. In Figure 5.11, the RF behavior of the new design is compared for the beam misalignment of 0.2 mm and 0.3 mm. The beam energy is linearly increased from 39 keV to 74 keV in the steps of 0.5 keV. The simulation setup for macro-electron based misalignment analysis has been already discussed in section 4.1. From the results, it is clearly observed that, the excitation is not stable with beam misalignment of 0.3 mm. After investigating the intermediate value, the stable operation can be achieved with the beam misalignment up

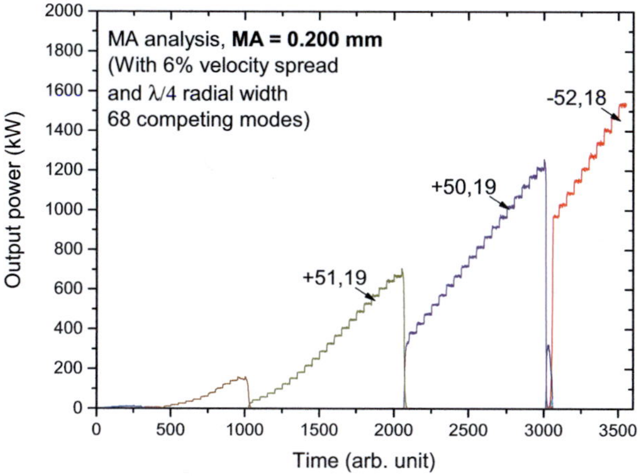

(a)

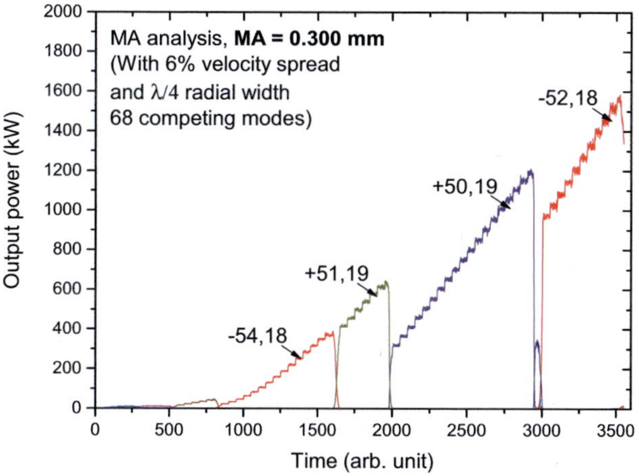

(b)

Figure 5.11: Comparison of gyrotron behavior at electron beam misalignment (MA) of 0.2 mm (top) and 0.3 mm (bottom). It is not possible to get stable output with beam misalignment of 0.3 mm.

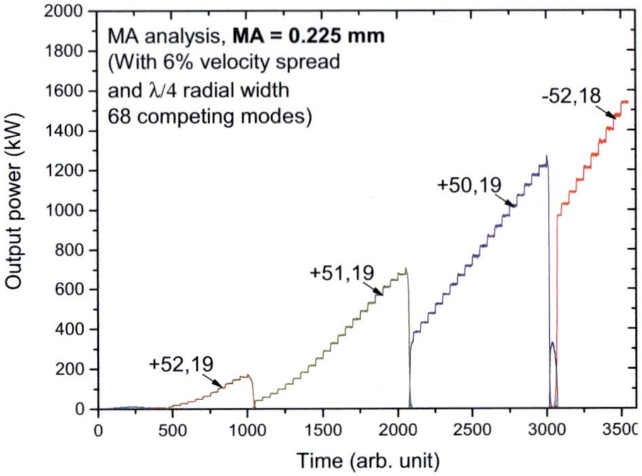

Figure 5.12: Result of multi-mode analysis considering electron beam misalignment of 0.225 mm. In new 1.5 MW design, the beam misalignment tolerance is reduced to 0.225 mm.

to 0.225 mm (see Figure 5.12). When only the radial width is considered in the analysis, this limit is extended to 0.250 mm. So, for the accurate estimation of beam misalignment tolerance, inclusion of velocity spread is a prerequisite. As compared to the standard design with $TE_{-43,15}$ mode, the beam misalignment tolerance reduces from 0.3 mm (D/λ = 0.24) to 0.225 mm (D/λ = 0.18).

6 Advanced start-up scenarios

In the previous chapter, the eigenvalue limit for a 236 GHz hollow-cavity gyrotron has been estimated using suggested approaches. Assuming a diode-type start-up scenario, new design has been proposed for the 1.5 MW output power. In this chapter, the possible solutions are investigated to further increase eigenvalue limit with the great focus on triode-type start-up scenario. The performance limits of 236 GHz hollow cavity gyrotron are also estimated in the wide range of eigenvalues.

Short interaction section length and eigenvalue limit

For a 236 GHz hollow-cavity design, considering standard cavity with 12 mm of interaction section length (L_2) and diode start-up scenario, the stable gyrotron operation is possible up to the mode eigenvalue 125. As discussed in section 2.2 and 5.7, the mode stability is dependent on the cavity interaction section length. The effective interaction length is reduced with the short cavity, which eventually reduces the quality factor (Q_{diff}) and improves the mode stability.

For the operating mode DM7 (eigenvalue $\sim$ 135) (refer Table 5.2), an RF-behavior of the gyrotron is investigated with the cavity interaction section length of 11 mm (L_2/λ = 8.65). All other physical parameters are kept same as per Table 5.7. The multi-mode time-dependent self-consistent scenario with the modified cavity length is presented in Figure 6.1. The result supports the stable excitation of mode $TE_{-56,20}$ with reduced interaction section length, which is not the case with the interaction section length of 12 mm (see Figure 5.6). Without including the effects of the realistic beam parameters, total output power of 1750 kW can be achieved with the interaction efficiency of 34 %.

The effects of interaction section length on mode stability is studied in [SALL92] [LA90] and the region for stable gyrotron operation is identified. With the short-interaction section length, output power and mode stability are improved, but with the reduced efficiency. To control overall plant efficiency, the minimum interaction efficiency of > 35 % is necessary for the DEMO gyrotrons. It is not possible to fulfill interaction efficiency requirements with the short interaction length. Therefore, for a 236 GHz DEMO gyrotron, this approach is not suitable to further extend eigenvalue limit. Up to now, only diode-type start-up scenario has been considered in this work.

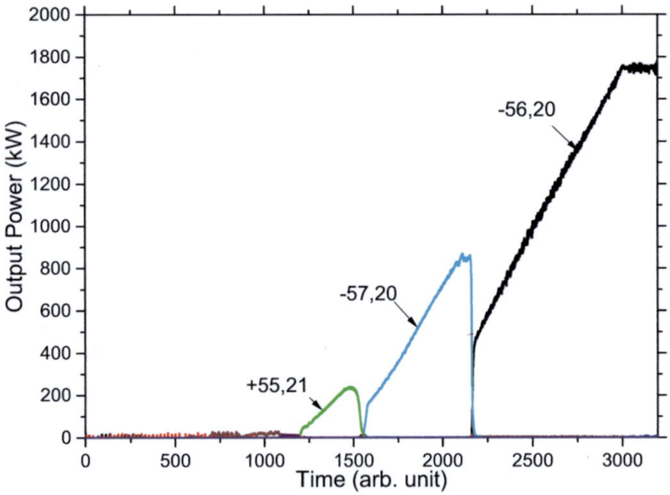

Figure 6.1: Multi-mode start-up scenario with DM7 ($TE_{-56,20}$) mode. Using the short interaction length (L_2) of 11 mm, it is possible to excite higher order mode with eigenvalue 135, which is not feasible with the standard cavity design (L_2 = 12 mm).

The influence of advanced triode-type start-up scenario on mode competition and eigenvalue limit will be studied in the next section.

6.1 Triode-type start-up scenario

The performance of gyrotrons having dense mode spectra are greatly dependent on the temporal growth of beam parameters. The carefully chosen start-up scenario is necessary to suppress the parasitic mode oscillations and to improve the mode stability of the desired mode. Investigations on the influence of start-up on the cavity mode have been presented in [BJ87] [LA90] [DS92] [KT87] [NVA^{+}08] [TSG^{+}13]. The diode-type start-up conditions have been considered in the previous analysis of this work (chapter 2 to 5) and RF behavior of gyrotron is determined. This section deals with the influence of triode-type start-up scenario on the mode competition control and eigenvalue limit.

The gyrotron gun designs are classified into two main types: (a) standard diode-type gun and (b) the triode-type gun. The schematic of the diode-type and triode-type gun is shown in Figure 1.4. In case of diode type gun, the value of beam energy (V_b) and the velocity ratio (α) is determined by the cathode voltage (V_c). With cathode voltage, the beam energy is increased to the nominal value. During start-up, the velocity ratio and beam current vary according to the adiabatic approximation [Che74]. Along the start-up, it is not possible to control the beam energy and velocity ratio independently, which leads to the excitation of high frequency parasitic modes before the excitation of the desired operating mode. In the triode-type gun, the individual control of beam energy and velocity ratio can be achieved using the additional modulating anode voltage (V_M), which allows different start-up scenarios to reach a nominal value of beam parameters [KNSA10].

The experimental verification of triode-type start-up scenario is presented in [SKT+07] [WTA+95][KOK+11] [IOK+16]. In [WTA+95], for an 118 GHz gyrotron, the mode competition between $TE_{-22,6}$ and $TE_{+19,7}$ mode has been studied and the effect of start-up scenario on mode stability has been verified. The triode-type start-up also facilitates the multi-frequency gyrotron operation by selecting various operating modes at different frequencies with an appropriate velocity ratio [KOK+11] [IOK+16]. The various triode-type start-up configurations have been investigated numerically in [WTTA94].

At given beam energy, the corresponding modulating anode voltage to achieve the specific value of the velocity-ratio (α) can be calculated using [WTTA94],

$$V_{\mathrm{M}} = \frac{m_{\mathrm{e}}c^2}{e}\frac{\ln(1+D_{\mathrm{F}}k)}{\ln(1+2k)} \tag{6.1a}$$

$$\left\{\left[1+\frac{4}{k^2}\left(\frac{1+k}{1+2k}\right)^2\left(\frac{\gamma^2-1}{R_{\mathrm{c}}^2(\cos\varphi_{\mathrm{c}})^2}\right)\left(\frac{\alpha_0^2}{\alpha_0^2+1}\right)\right]^2-1\right\}.$$

Where,

$$k = \frac{1}{\sqrt{R_{\mathrm{g}}^2-1}},\ R_{\mathrm{c}} = \frac{r_{\mathrm{ct}}}{r_{\mathrm{L}}} \tag{6.1b}$$

$$R_{\mathrm{g}} = \frac{r_{\mathrm{beam}}}{r_{\mathrm{L}}},\ D_{\mathrm{F}} = \frac{\cos\varphi_{\mathrm{c}}}{k}\frac{d_{\mathrm{ac}}}{r_{\mathrm{ct}}} \tag{6.1c}$$

Here, d_{ac} = cathode-anode separation distance, r_{ct} = average cathode radius, γ = relativistic factor, ϕ_{c} = cathode slant radius. Along the start-up, it is possible to have the full-control of velocity ratio (α) with proper selection of anode voltage.

The triode start-up scenario can be performed in two possible ways: (a) for particular beam energy, the equivalent anode voltage (V_M) is calculated using equation 6.1a to generate a beam with desired pitch factor. As a first step, the beam energy is increased to the optimal value by increasing cathode voltage (V_c). Once the desired beam energy is attained, the anode voltage is increased from zero to V_M. (b) In the second approach, the desired pitch factor variation along the start-up is achieved by introducing proper delay between the rise of the cathode voltage and the modulating anode voltage [SNAV08].

For example, in Figure 6.2, the beam parameters during the diode and triode start-up scenarios are compared, in which the beam energy is varied from 40 keV to 80 keV. In case of diode start-up, the beam energy increases linearly with the adiabatic change in the beam current and velocity ratio, while in the triode start-up case, the velocity ratio is varied in following three stages,

- Stage I: beam electron energy: 40 keV – 65 KeV, velocity ratio: 0.5 (constant)
- Stage II: beam electron energy: 65 keV – 66 KeV, velocity ratio: 0.5 – 1.15 (linear rise)
- Stage III: beam electron energy: 66 keV – 80 KeV, velocity ratio: 1.15 – 1.25 (linear variation)

For this analysis, the region, where the pitch-factor is varied abruptly (stage II) is termed as the "Transition phase". In Figure 6.2, the beam energy of 65 keV to 66 keV is chosen as the "Transition Phase", where the pitch-factor changes from 0.5 to 1.15. The effects of a triode start-up scenario on the mode competition and the selection criteria of operating parameters are discussed in the following sections.

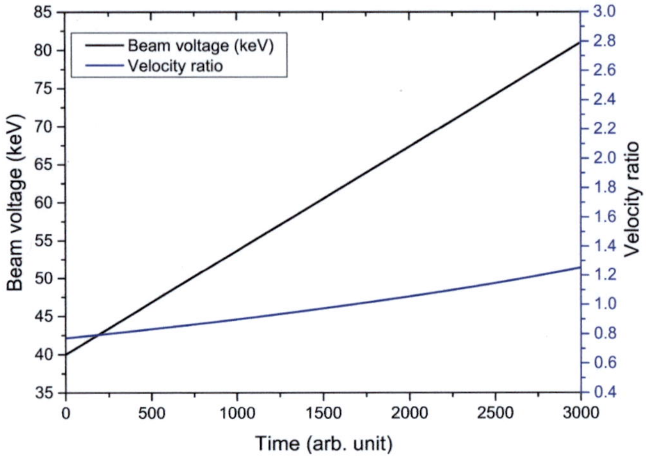

(a)

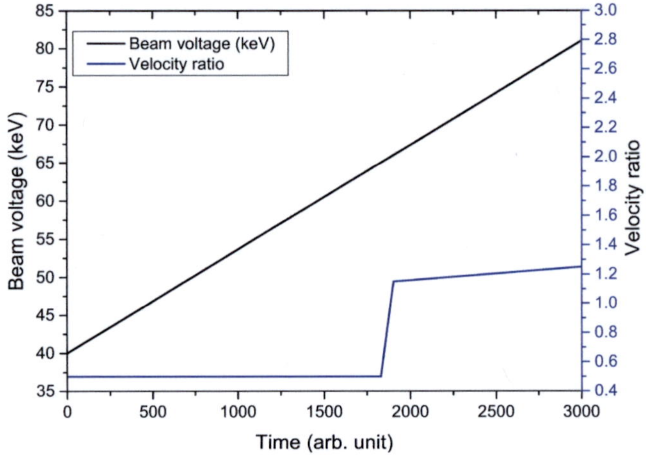

(b)

Figure 6.2: Comparison of the beam-parameter variation for (a) diode-type and (b) triode-type start-up scenario. In diode-type start-up, velocity ratio (α) follow adiabatic approximation, while, in the case of triode-type start-up, the velocity ratio can be modified using modulating anode voltage.

6.1.1 Effect on mode competition

The influence of triode start-up scenario on mode competition is studied with the help of starting current curves. The detailed discussion of the starting current calculations has been covered in section 5.6.1. In this part, the DM9 case of Table 5.2 (operating mode: $TE_{-59,21}$, eigenvalue $\sim$ 145) is selected for this analysis. The selected cavity mid-section radius is 29.14 mm and the calculated beam radius is 12.37 mm. The other physical parameters are same as Table 5.7. From the results of the eigenvalue limit analysis presented in chapter 5, it is proved that the stable mode excitation of $TE_{-59,21}$ is not possible using diode-type start-up scenario.

In Figure 6.3, the starting current curves of the diode type and triode type start-up scenario have been compared to the variation of beam energy from 50 keV to 80 keV. The corresponding beam current has also been plotted for both the cases. In diode start-up, the cathode voltage increases linearly with adiabatic changes in the beam current and the pitch factor. As mentioned before, the triode type start-up scenario is divided into three stages: stage I: 50 keV to 65 keV (velocity ratio = 0.5, constant value), stage II: 65 keV to 66 keV (velocity ratio = 0.5 to 1.5, linear rise) and stage III: 66 keV to 80 keV (velocity ratio = 1.5 to 1.25, linear variation).

During the first stage, the low velocity-ratio shifts the starting current plots of the parasitic modes to very high value, which suppress the excitation of any mode during stage I. In the second phase of start-up, the velocity-ratio increases linearly from 0.5 to 1.15 and provide suitable conditions for mode excitation. At the time of transition phase, the mode having minimum starting current has the highest probability for excitation (in Figure 6.3, it is $TE_{-60,21}$). The selection of the transition phase is very crucial for triode start-up and the start-up scenarios can be easily modulated by selecting proper transition phase. The detailed analysis of the selection of transition phase and its effects on mode-excitation are presented in the next section. During stage III, the operating parameters attain their desired value of the steady-state operation.

The starting current of parasitic modes are very high during the stage I and stage II, which controls the mode competition during the start-up phase. From the results, it is clear that the triode type start-up modifies coupling scenario and suppress the excitation of parasitic modes during start-up. Using the multi-mode self-consistent simulations, the effects of triode start-up on mode competition control is further investigated in the next section.

6.1.2 Multi-mode start-up simulations

For the diode start-up scenario, eigenvalue limit of a 236 GHz DEMO gyrotron has been determined in chapter 5. The various generic approaches suggest the stable mode operation till eigenvalue of 125, which corresponds to the maximum possible output power of 1.5 MW. In this section, the effect of the triode-type start-up on mode competition and mode eigenvalue limit are studied.

An RF behavior of a 236 GHz hollow-cavity gyrotron with the operating mode $TE_{-59,21}$ (eigenvalue $\approx$ 145) is verified in Figure 6.4, for diode-type and triode-type start-up cases. The beam energy changes linearly from 40 keV to 80 keV. In total 84 neighboring modes are considered for the multi-mode analysis with realistic beam parameters (6 % (rms) velocity spread and $\lambda/4$ radial width). With the mode eigenvalue $\approx$ 145, it is not possible to get stable operation with $TE_{-59,21}$ mode. This result (Figure 6.4(a)) also supports the findings of eigenvalue limits from chapter 5.

In the case of the triode start-up, as discussed in section 6.1, the pitch-factor variation is considered in three stages. The transition phase is selected from

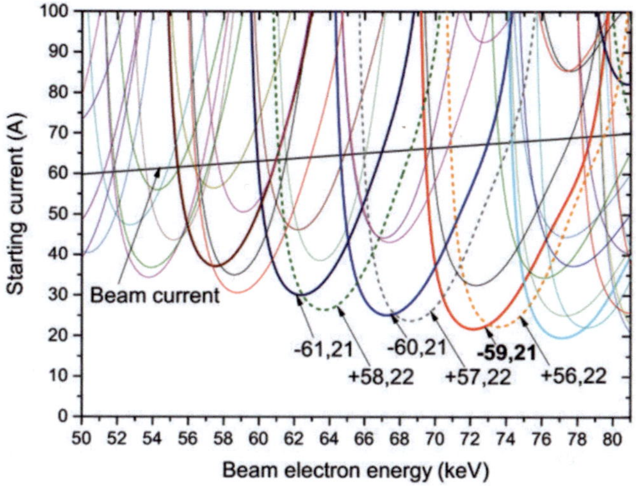

(a)

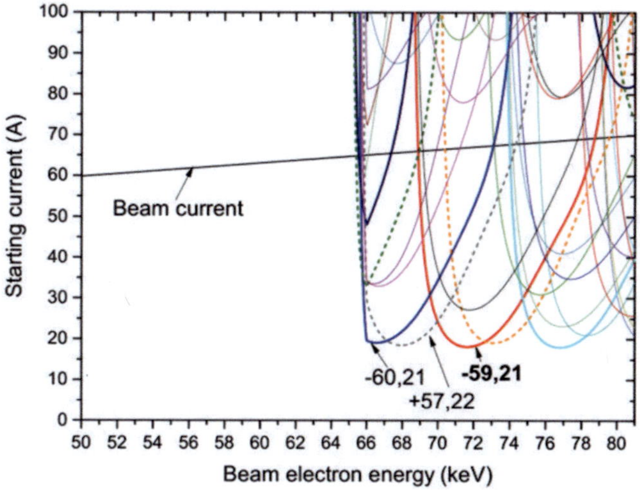

(b)

Figure 6.3: Starting current plots for the $TE_{-59,21}$ mode during (a) diode-type and (b) triode-type start-up scenario.

65 keV to 66 keV. The multi-mode simulation with the modified triode start-up scenario is presented in Figure 6.4 (b) and the transition phase is indicated with the "vertical blue dashed lines". Because of the low pitch factor, no mode excites during stage I (beam energy: 40 keV to 65 keV). During the transition phase, as pitch factor increases to 1.15, the counter rotating $TE_{-60,21}$ mode excites, which lead to the excitation of the $TE_{-59,21}$ mode in the final stage.

Unlike the diode start-up case, the mode competition is controlled in the triode start-up scenario and a stable operation is possible with the $TE_{-59,21}$ mode. The reduced mode coupling of the neighboring modes during stage I, is helpful to select particular mode and to excite desired operating mode. The starting current scenario presented in Figure 6.3 (b), also suggest the excitation of $TE_{-60,21}$ after stage II. In this case, the position of transition phase plays a significant role in mode excitation during the start-up and it is specifically chosen to excite same series mode ($TE_{-60,21}$), before operating mode $TE_{-59,21}$.

6.1.3 Mode selectivity using triode start-up

The detailed analysis of the transition phase position on start-up scenario is presented in this section. During the gyrotron start-up, the excitation of high-frequency neighboring modes can not be controlled in the case of diode start-up. In such cases, beam energy increases from the noise level to reach up to the final optimized point and there is no control on the excitation of the start-up modes.

The proper position of the transition phase and specific variation of the velocity-ratio (α) is a prerequisite for the successful triode start-up scenario. Based on the position of the transition phase, the first mode excites during

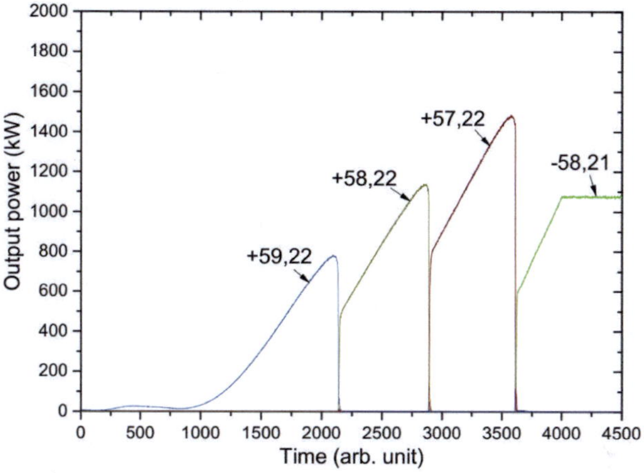

(a)

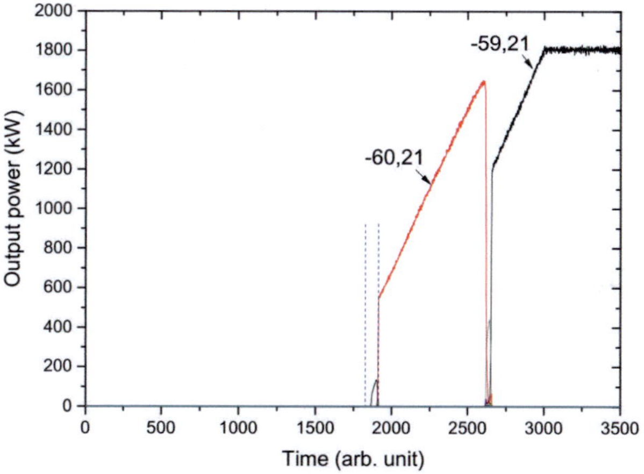

(b)

Figure 6.4: (a) Diode-type and (b) triode-type start-up simulation with the operating mode $TE_{-59,21}$ at mode eigenvalue $\approx$ 145. Due to very high mode competition, it is not possible to excite desired mode using diode start-up. In the case of triode start-up, the controlled mode competition lead to stable mode excitation.

the start-up, which further excites the final operating mode. After careful investigation of starting current plots and mode coupling, the position of the transition phase can be finalized. For example, in case of DM9, the starting current for the $TE_{-60,21}$ mode (see Figure 6.3) is minimum at the beam energy around 66 keV. So, the selection of transition phase at around 66 keV leads to the excitation of $TE_{-60,21}$. In Figure 6.4 (b), in total two modes (1 neighboring mode ($TE_{-60,21}$) + main mode($TE_{-59,21}$)) are excited using triode start-up, by selecting transition phase between 65 keV and 66 keV. The mode selectivity using triode start-up is further studied with the help of following cases.

Case 1: Two start-up modes

The main aim of this case is to excite three modes (2 parasitic modes $TE_{-61,21}$ and $TE_{-60,21}$ with the main mode $TE_{-59,21}$) by considering proper transition phase and pitch-factor variation. The starting current of mode $TE_{-61,21}$ is minimum between the beam energy of 60 keV and 62 keV (see Figure 6.3). The excitation of the $TE_{-61,21}$ mode is favored, if the transition phase is selected in this region. In this analysis, the pitch-factor (α) is varied from 0.5 to 1.15 with the beam energy selected between 61 - 62 keV (stage II). It is constant (α = 0.5) between the beam energy of 40 keV to 61 keV (stage I) and increases linearly from 1.15 to 1.25 between the beam energy of 62 keV to 80 keV. The multi-mode start-up simulations with these settings are presented in Figure 6.5 (a). These results support stable operation of $TE_{-59,21}$ with initial excitation of $TE_{-61,21}$ and $TE_{-60,21}$, which also confirm the mode selectivity with proper selection of the transition phase.

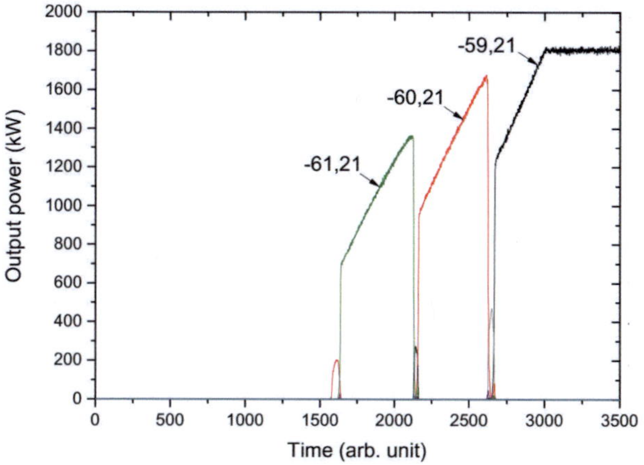

(a)

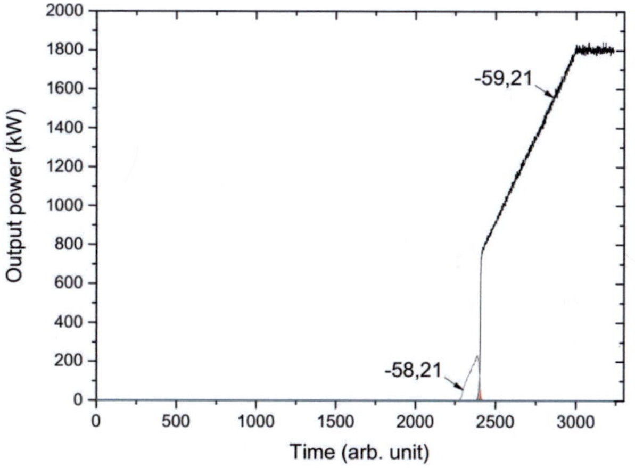

(b)

Figure 6.5: With the suitable alpha transition region, the mode selectivity can be achieved using triode start-up scenario. (a) The alpha transition phase is selected between 61 keV to 62 keV, which leads to the two starting mode excitation before main mode, while in (b), the transition phase is selected at 70 keV to 73 keV, which supports the direct mode excitation.

Case 2: Direct excitation

The possibilities of mode selectivity using triode start-up are further verified for the direct excitation of the main mode without any previous modes. After studying starting current curves, the transition region (stage II) is selected between 70 keV to 73 keV. The direct excitation of main mode $TE_{-59,21}$ is possible with selected operating condition. The direct operating mode excitation is not possible using the diode start-up, but, with the triode start-up, it is possible with the proper selection of the pitch-factor variation.

The mode selectivity has been also verified with the counter-rotating modes. Considering the limitations from the gyrotron power-supply, slow variation of pitch-factor (long transition phase, stage II) has been also verified and suggests the similar mode selectivity and mode competition control.

As discussed in [WTTA94], the slope of pitch-factor variation (start and end values of velocity ratio variation in stage II) is critical for the desired mode excitation and greatly affects the start-up scenario [WTA+95]. To confirm desired start-up condition, the values of the velocity-ratio for stage II are selected using velocity spread (α) vs beam energy plots.

Though, it is based on the fixed-field approximation, these plots provide a first estimation of the pitch-factor values for the triode start-up scenario. The contours of starting current plots of individual modes are plotted in alpha vs beam energy plan (see Figure 6.6). To generate these contours, within the selected beam energy, the starting current for the cavity modes are calculated at various pitch-factors and the contour of the same starting current values are plotted. These contours represent the minimum pitch-factor requirement to excite particular mode at the specific beam energy and beam current. In Figure 6.6, the contours are plotted for the main mode $TE_{-59,21}$ and the important competing modes for the starting current of 70 A. Within

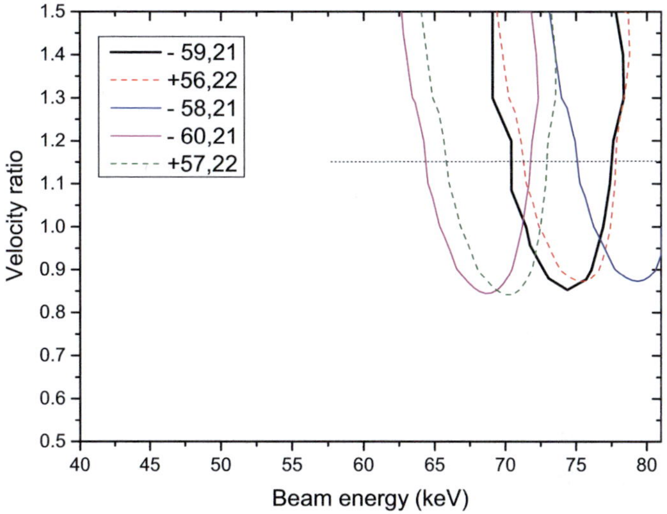

Figure 6.6: Velocity ratio (α) vs beam energy contours with starting current value of 70 A.

the selected range of the beam energy, the starting current values are calculated at various velocity ratio and the contours of velocity ratio (α) vs beam energy are obtained. From the result, it is confirmed that the selected value of a pitch-factor in stage I ($\alpha = 0.5$) is very low to excite any mode. To excite the desired mode during triode start-up, the slope of a pitch factor variation is selected using Figure 6.6. The pitch factor of 1.15 is sufficient to excite all the modes plotted in Figure 6.6. By controlling the position of the transition phase and slope of pitch-factor variation, various start-up scenario can be possible with a triode-type gun.

In chapter 2, the stable output power of 920 kW has been suggested with the selected operating mode $TE_{-43,15}$ (eigenvalue $\sim$ 103). As concluded from the discussion of section 4.2.3, the mode eigenvalue limit for the stable gyrotron operation is 125 with the standard diode-type start-up scenario, which corresponds to the output power of 1.5 MW. In addition to the possibilities of mode selectivity, the triode-type start-up scenario is very effective

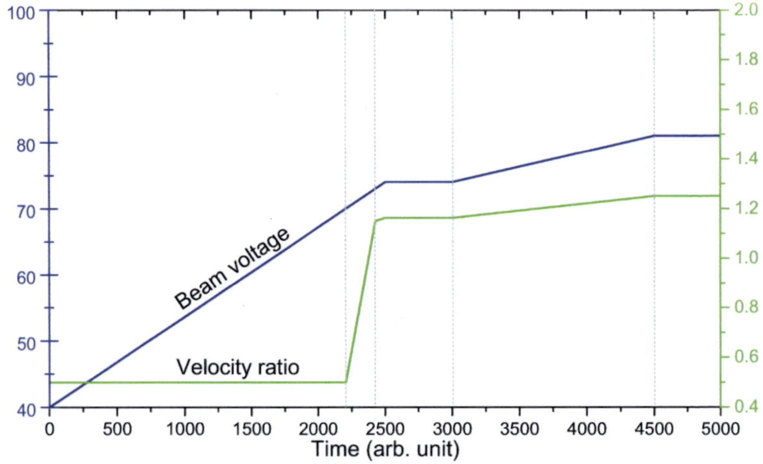

Figure 6.7: Beam parameters for the new hybrid start-up scenario. In this case, desired mode is excited in soft excitation region and then, as a next step, beam parameters are tuned for the hard excitation case to achieved optimum efficiency.

to control the mode competition and mode eigenvalue limit is numerically extended to 145 (DM9, $TE_{-59,21}$ mode). Taking into account the lifetime of gyrotron gun, the beam current more than 70 A is not desirable for fusion gyrotrons [IFMC10]. Because of the limitations of beam current, the modes having eigenvalue greater than 145 are not considered for the analysis. It should be also noted that the effects of beam misalignment and space-charge neutralization are critical for high order mode (eigenvalue $\sim$ 145), which demands rigorous tolerance analysis before the final mode selection.

6.1.4 Hybrid start-up condition

In the case of gyrotron with a uniform magnetic field, the maximum efficiency operating points are located in the hard-excitation region [Nus04] [WTTA94]. As compared to the soft-excitation region, the interaction ef-

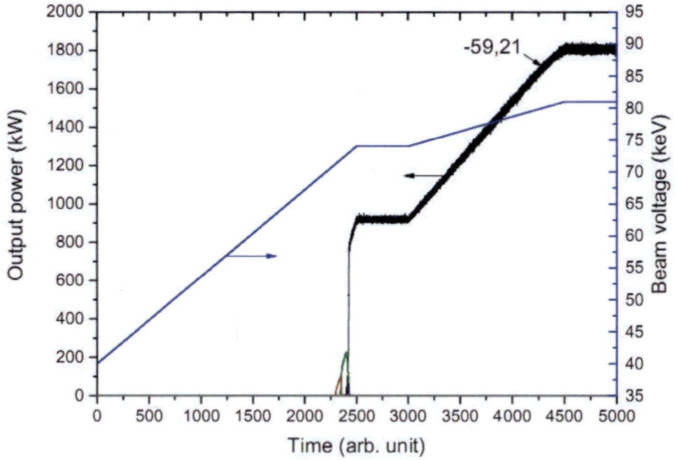

Figure 6.8: Hybrid-type start-up simulation for $TE_{-59,21}$ mode (eigenvalue $\sim$ 145).

ficiency shall be increased more than 5 % in the hard-excitation operating point. In this section, the main difference between these two regimes are discussed briefly in the context of beam current and starting current plots. In the soft-excitation regime, the value of beam current is higher than the value of the starting current. The self-excitation conditions are fulfilled within this operating range. From the starting current plots presented in Figure 6.4, the soft-excitation regime for the operating mode $TE_{-59,21}$ is between the beam energy of 69 keV to 78 keV.

In the regime of hard-excitation, the value of beam current is less than the value of starting current. So, the stable operation in this regime is only possible, when initial mode excitation reaches a certain threshold value. With the help of a suitable start-up scenario, one should pass the soft-excitation region by modulating gyrotron operating parameters, in which excitation of the mode initiates and will continue until the hard excitation point is reached. The effects of various possible start-up scenarios are studied in

[WTTA94] and proved that the mode generation is independent of the start-up path in the soft-excitation regime, while to reach the hard-excitation point, the specific start-up conditions are a prerequisite.

To get advantages of both the cases, the "Hybrid-type mode excitation" approach is suggested. In this approach, initially using the triode start-up, the operating mode is excited in the soft-excitation regime and stabilized the mode. As a next step, the operating parameters are tuned to achieve hard-excitation operating point. The unique advantages of this approach include controlled mode competition due to the triode start-up, stable operation independent of start-up path due to the initial soft-excitation and high-efficiency operation as the final operation is in the hard-excitation regime.

The example of hybrid-type mode excitation scenario with the $TE_{-59,21}$ mode is presented in Figure 6.7 and 6.8. The velocity ratio is increased from 0.5 to 1.15 between the beam energy of 70 keV to 73 keV. Based on the starting current plots, the detailed discussion about this selection is already presented in section 6.1.3. At the beam energy of 74 keV, the operation is stabilized in the soft-excitation region and the hard-excitation point can be achieved by slowly changing the operating parameters. During the soft-excitation, the stable output of around 1 MW can be achieved using the multi-mode time-dependent simulations, which is increased to 1.8 MW in hard-excitation regime.

6.2 Output power and efficiency limits

In addition to the mode eigenvalue, the performance of the gyrotron is highly dependent on the interaction section length (L_2) [LA90]. After investigating mode eigenvalue limit, it is also important to have information about the influence of interaction section length at different eigenvalues. The main focus of this analysis is to find output power and efficiency limits at dif-

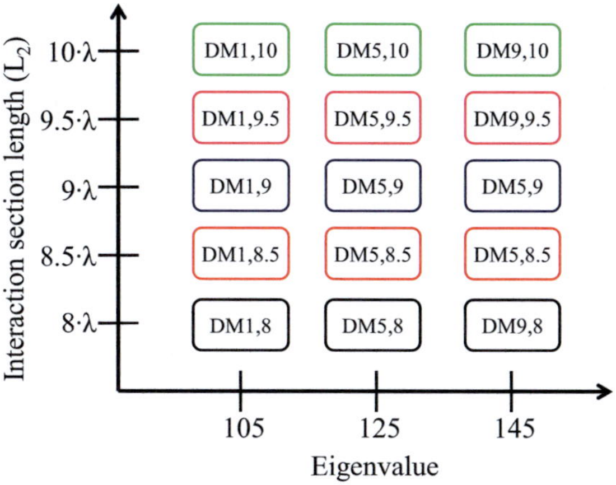

Figure 6.9: Notation used in section 6.2, to represent individual case.

ferent mode eigenvalues with emphasis on mode stability. This analysis is only suitable for the first order estimation of gyrotron behavior at particular eigenvalue and interaction length. The final case specific optimization is necessary as per the requirements.

The three modes having different eigenvalues are selected for analysis: $TE_{-44,15}$, $TE_{-56,20}$ (eigenvalue limit with diode start-up) and $TE_{-59,21}$ (eigenvalue limit with triode start-up). The different interaction lengths between $8 \cdot \lambda$ to $10 \cdot \lambda$ are considered. In Figure 6.9, the notation used in each case is presented. The self-consistent time-dependent simulations have been performed with various interaction lengths at particular eigenvalue and effects of mid-section length on mode stability are studied. Since the mode eigenvalue limit for diode start-up is 125, the diode-type start-up scenario is used for DM1 ($TE_{-44,15}$) and DM5 ($TE_{-56,20}$) modes, while triode start-up is used for the analysis of DM9 ($TE_{-59,21}$) mode. Similar to the previous analysis, the cavity wall-loading limitation of 2 kW/cm^2 is considered and the operating parameters are finalized accordingly. For the selected interaction length interval ($8 \cdot \lambda$ (DM1,8) to $10 \cdot \lambda$ (DM1,10)), the operating parameters and gyrotron performance with the operating mode $TE_{-44,15}$ is presented in Figure 6.10. With the $TE_{-44,15}$ mode (eigenvalue $\sim$ 105), the stable gyrotron operation is possible within the entire selected interaction section range. Depending on the interaction length, the output power varies within the range of 1 MW – 1.2 MW.

The results with mode $TE_{-52,18}$ (eigenvalue $\sim$ 125) are presented in Figure 6.11. Because of the critical mode competition, with interaction section length of 12.7 mm (case DM5,10 of Figure 6.9), it is not possible to get stable operation (Q $\sim$ 1726). So, the maximum interaction section length for the stable operation is limited to 12.07 mm (L_2/λ = 9.5). The output power and interaction efficiency range with mode eigenvalue close to 125 are 1.5 MW – 1.6 MW and 33 % - 38 %, respectively.

In the case of mode eigenvalue 145 ($TE_{-59,21}$ mode), the optimal interaction length window further shrinks to only 8.5 $\cdot\lambda$ (DM9,8.5) to 9.5 $\cdot\lambda$ (DM9,9.5) (see Figure 6.12). With interaction length L_2 = 10.16 (L_2/λ = 8), the beam current requirement increases to more than 80 A for efficiency higher than 30 %, which is not recommended considering a lifetime of gun and voltage-depression. Similar to DM5,10 case, the quality factor of cavity and mode competition is critically high with the interaction section length of 12.7 mm (DM9,10).

The results of this analysis are compiled in Figure 6.13, in which the output power and efficiency range at the different eigenvalues are presented. For the selected eigenvalue ranges of 105 to 145, the interaction length of 11.5 mm (L_2/λ = 9) is the best choice for the stable operation with sufficiently good interaction efficiency and controlled mode competition. It should be also noted that the consideration of the realistic beam parameters (velocity spread, radial width etc.) shall marginally reduce the estimated output power and efficiency. Considering the cavity wall-loading of

2 kW/cm^2, the suitable operating parameter windows are also suggested for the different combination of the interaction length (L_2) and mode eigenvalues (See Figure 6.14). Using these plots, the performance and the operating parameters of a 236 GHz gyrotron can be easily estimated for the intermediate modes (or eigenvalues).

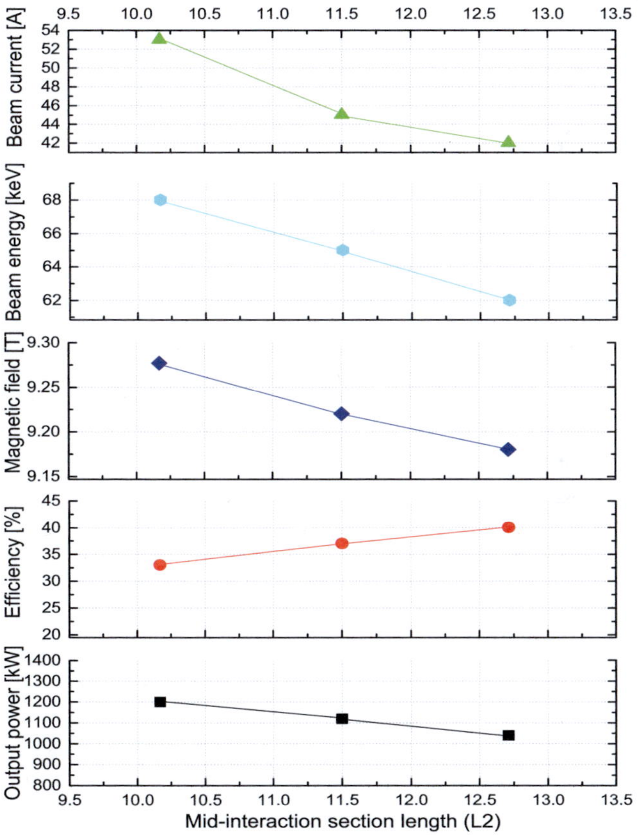

Figure 6.10: For the operating mode $TE_{-44,15}$ (eigenvalue $\sim$ 105), influence of interaction section length on the operating parameters and gyrotron performance.

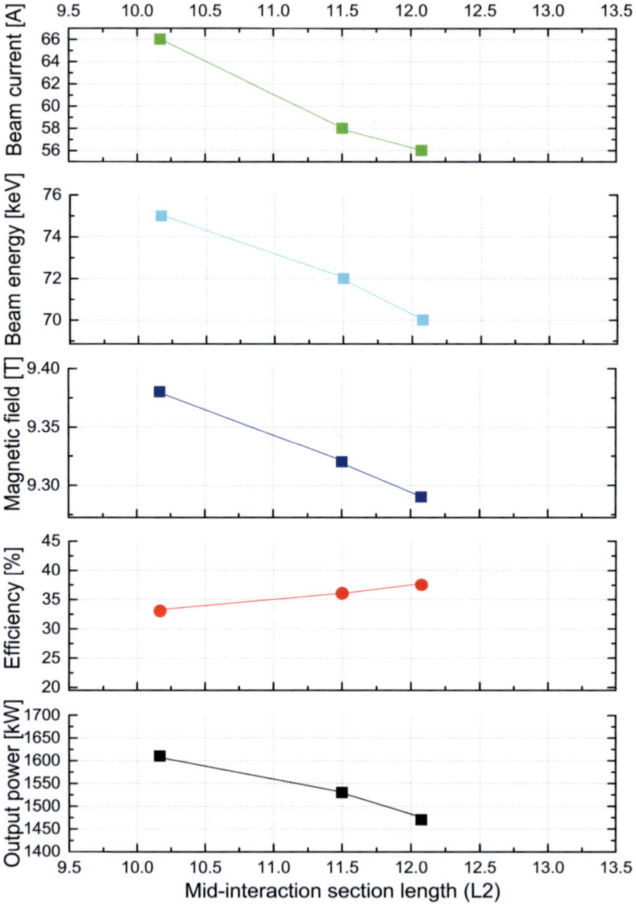

Figure 6.11: The effects of interaction section length on the gyrotron performance for the operating mode $TE_{-52,18}$ (eigenvalue $\sim$ 125).

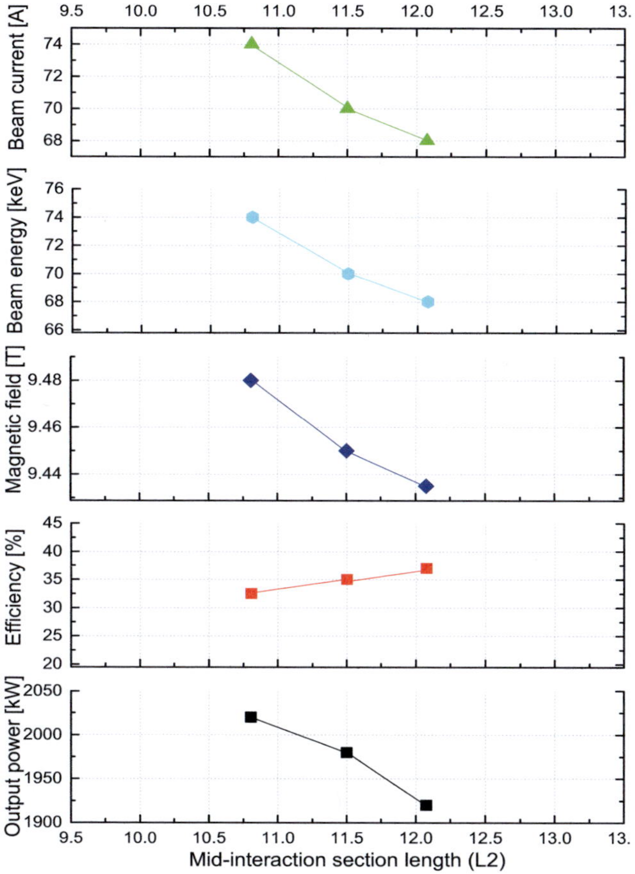

Figure 6.12: Influence of interaction section length on the operating parameters and gyrotron performance with operating mode $TE_{-59,21}$ (eigenvalue $\sim$ 145).

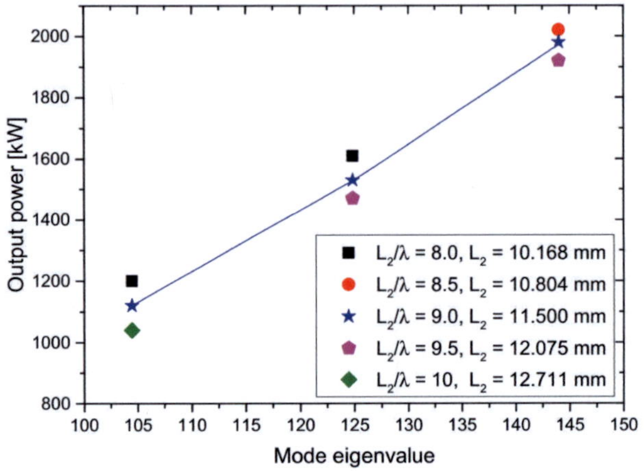

(a)

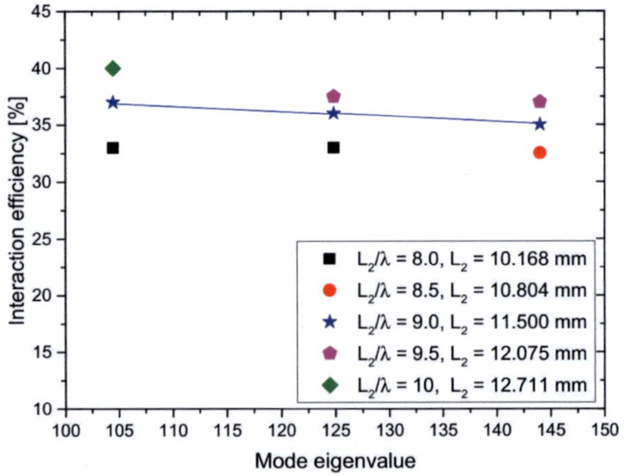

(b)

Figure 6.13: (a) Output power and (b) efficiency for the different eigenvalues between 105 to 145.

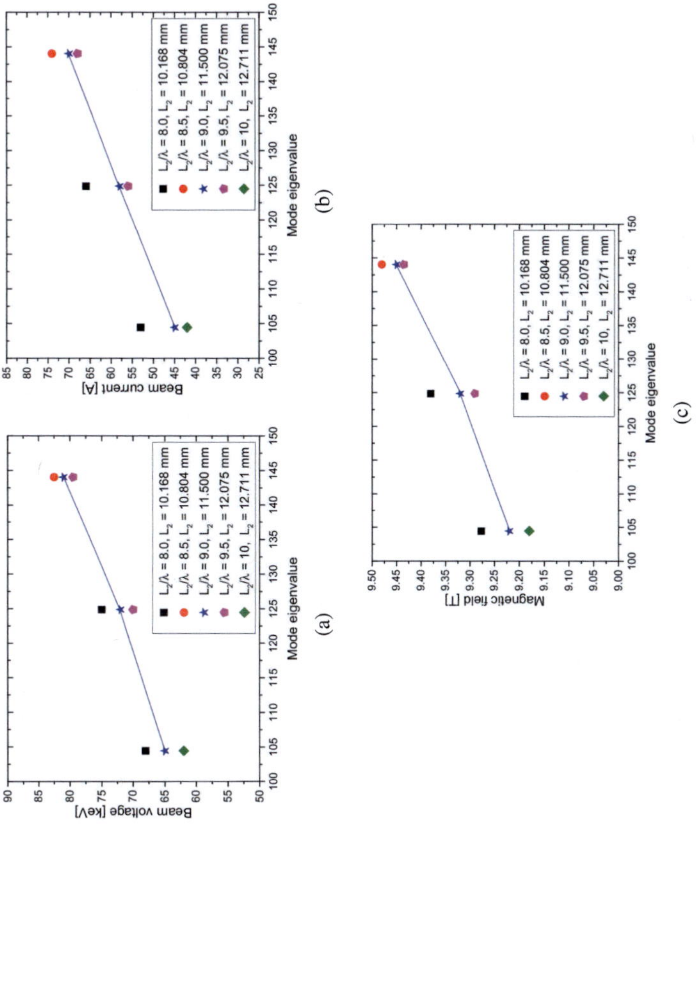

Figure 6.14: (a) Beam voltage (b) beam current and (c) magnetic field values at eigenvalue between 105 to 145.

7 Conclusions and outlook

After ITER, the first prototype of a fusion power plant termed as DEMO is foreseen and the applications of ECRH&CD systems have been planned at various stages of the operation. In this work, the physical design studies and the analysis of an advanced DEMO-compatible hollow cavity gyrotron have been performed with the focus on operation reliability, realistic interaction simulations and mode competition control. For the proposed design, the possibilities of multi-frequency, multi-purpose applications and fast-frequency step-tunability have also been verified. In addition, the operational limits of the DEMO gyrotron have been investigated to facilitate the highest possible output power per tube with desired mode stability.

The mode selection criteria for high frequency (> 200 GHz) fusion gyrotrons have been discussed and the selection of high order modes (eigenvalue > 100) has been suggested to satisfy the power requirements. At this frequency range, the mode spectra are identical and the selection of operating modes is based on multi-frequency operation and frequency step-tunability. Considering the RF-window thickness of 1.861 mm and low-frequency operation at 170 GHz (frequency of ITER gyrotron), the operating frequency of 236 GHz was selected for the hollow cavity design. A systematic cavity design approach has been proposed to finalize the physical parameters of the cavity. Using this, the optimal interaction section design was suggested for the $TE_{-43,15}$ mode at 236 GHz. Based on DEMO design goals and present technical constraints, the operating parameters were selected for efficient beam-wave interaction. An RF-behavior of the cavity was also analyzed using the multi-mode, time-dependent simulations. For realistic simulations, the effects of beam velocity spread and the radial width have been considered along with the frequency and temperature dependent Glidcop conductivity. The influence of the beam radius on gyrotron performance has been investigated and the suitable value of the beam radius has been selected for controlled mode competition. Despite of the consideration of the large number of neighboring modes (99 modes), the results suggest the stable excitation of $TE_{-43,15}$ mode with the output power of 920 kW and interaction efficiency of 36 %. The stability of operation was also verified for the wide range of operating parameters.

Besides the high output power and improved interaction efficiency, the fast-frequency step-tunability is needed for NTM stabilization using a fixed launcher. The suitable operating modes for frequency step-tunability should have nearly identical caustic radius, which allows an efficient mode conversion with the same quasi-optical launcher design. Within $\pm$ 10 GHz frequency range, the possibility of fast-frequency tunability in 2 - 3 GHz steps were numerically demonstrated using the magnetic field tuning. The speed of frequency tunability is limited by the rate of change of the magnetic field inside the cavity. In order to facilitate the multi-purpose applications, the proposed design additionally supports the multi-frequency operation (within a few minutes) at 170 GHz, 203 GHz, 236 GHz and 268 GHz, with the operating modes of $TE_{-31,11}$, $TE_{-37,13}$, $TE_{-43,15}$ and $TE_{-49,17}$, respectively. It should be noted that, the multi-frequency operation is possible using single-disk CVD-diamond window, but for fast-frequency step-tunability the use of the broadband Brewster window is a prerequisite. The hybrid-type launcher was designed using the in-house code "TWLDO" and more than 96 % of the fundamental Gaussian mode content was achieved with the selected modes for multi-frequency operations. The average Gaussian content was more than 95 % for the modes for fast frequency step-tunability.

Due to high-order operating mode (eigenvalue > 100), the effects of electron beam misalignment are critical at DEMO frequencies. The beam misalignment tolerance has been studied using a macro-electron based simulation approach. The results of the misalignment study support the stable gyrotron operation, till an axial misalignment of 0.3 mm (D/λ = 0.23). This was found to be comparable to the beam misalignment tolerance from the experiments of W7-X (140 GHz, 1 MW) gyrotron (D/λ = 0.25 - 0.3). Continuous wave operation of the DEMO gyrotrons has been planned to support the steady-state operation of the reactor. In chapter 4, the effects of space-charge neutralization have been studied to ensure CW operation of the DEMO gyrotron. Considering the variation of beam parameters during CW operation, a step-by-step simulation approach has been proposed for multi-mode start-up scenario. The output power of high frequency (> 200 GHz) gyrotrons are limited by the maximum allowable cavity wall-loading. For the specific cavity cooling systems and the operating frequency, the output power of the gyrotron can be improved by selecting even higher order modes, which allows a large cavity radius with reduced wall-loading. However, the mode competition increases proportionally. Using two different methods, the effects of mode competition were investigated for a 236 GHz hollow-cavity gyrotron design and the eigenvalue limits for the operating mode selection were determined. Moreover, the suggested approaches for the operational limit evaluation are generic and also valid for other gyrotron designs. Assuming diode start-up and realistic electron beam parameters, these methods confirm the stable operation with modes having eigenvalues of up to 125. Based on the suggested eigenvalue limit, a new cavity design and operating parameters were proposed, which corresponds to an output power of 1.5 MW using $TE_{-52,18}$ mode.

The possibilities of advanced start-up scenarios were studied to further extend the operational limits. As compared to the diode-type gun, the triode-type gun provides better control on the beam parameters (specially pitch factor), which promotes the controlled mode competition with mode selective operation. The gyrotron performance with the hybrid-type start-up scenario was verified, which suggested unique advantages such as: controlled mode competition due to the initial triode start-up, high efficiency in hard-excitation regime and stable operation which is independent of the selected start-up path.

Summarizing, a rigorous feasibility and tolerance analysis were performed for a hollow-cavity gyrotron design. This is a step forward towards the development of the next generation multi-frequency and frequency step-tunable DEMO-gyrotrons with the improved output power and interaction efficiency. The key parameters of a 236 GHz, 1 MW and 1.5 MW DEMO gyrotron designs are listed in Table 7.1.

During the next development phase of hollow-cavity DEMO gyrotrons, the activities for efficient, stable and high-power operation are briefly discussed below. In case of high frequency (f > 200 GHz) fusion gyrotrons, the modes having eigenvalue more than 100 are selected to control the wall-loading and to fulfill output power requirements. At such high eigenvalues, mode competition is critical. In chapter 6, the triode-type start-up scenarios were studied to control the mode competition. In addition to the triode start-up, alternative techniques such as injection locking [BDN16] [ZGD$^+$16], complex interaction section [ZY17] [LNY$^+$13] [MRR$^+$16] [BKS16] etc., should be investigated for improved mode stability. At IHM-KIT, a 10 T magnet system for DEMO gyrotrons is under discussion and the magnetic field profile along the gyrotron axis is not yet finalized. Using the realistic magnetic field profile, the designs proposed in this thesis can be additionally verified for After Cavity Interaction (ACI) and the unwanted parasitic oscillations in the beam tunnel region. The effect of space-charge neutralization shall also be included in the triode-type start-up scenario for realistic long-pulse simulations.

Table 7.1: Physical parameters of the cavity, operating parameters and performance comparison of the suggested 1 MW and 1.5 MW, 236 GHz DEMO gyrotron design. (maximum cavity wall-loading: 2 kW/cm^2, perpendicular velocity spread: 6 % and beam radial width: $\lambda/4$)

Parameters	1 MW design	1.5 MW design
Operating mode	$TE_{-43,15}$	$TE_{-52,18}$
Eigenvalue	103.21	124.87
Physical parameters of cavity		
Cavity radius (mm)	20.88	25.26
$L_1/L_2/L_3$ (mm)	16/12/16	16/11.5/16
D_1/D_2 (mm)	2/2.5	2/2.5
$\theta_1/\theta_2/\theta_3$	2.5°/0°/2°	2.5°/0°/2°
Operating parameters		
Beam radius (mm)	9.06	10.93
Beam current (A)	43	63
Beam energy (keV)	61	74
Magnetic field (T)	9.165	9.355
Performance with realistic beam parameters		
Output power (kW)	920	1530
Efficiency (%)	36	34
Beam misalignment tolerance (mm)	0.300	0.225

Considering the present technical limitations, the future high power gyrotrons demand advanced MIGs with high current density, improved lifetime and minimum spread. New disruptive gun technologies are to be developed to meet future gun requirements such as Inverse Magnetron Injection Gun (IMIG) [RPG+16] [RPR+15] and Controlled-Porosity Reservoir

(CPR) cathode [IFSW05] [IFMC10]. In section 4.1, the effects of beam misalignment have been studied using realistic macro-electron beam based approach. It is observed that the performance of DEMO gyrotrons is very sensitive to beam misalignment. The stable operation is only possible with beam misalignment less than 0.3 mm. The sophisticated beam misalignment control system needs to be developed for the precise beam placement. The gyrotron output power is limited by the cavity cooling capacity. With an improved cavity cooling system, it is possible to increase the maximum wall-loading capability, which eventually enhances the power handling capacity of the interaction section. Additionally, the new concepts of multistage depressed collector are to be developed to satisfy the total efficiency requirement of the DEMO gyrotrons ($\eta_{\text{total}} > 60$ %).

A Appendix

A.1 Calculation of Glidcop conductivity

Glidcop is a copper alloy with enhanced thermal properties. In this part, the formulation for the realistic Glidcop conductivity calculation is discussed. The absorption losses of the Glidcop have been measured at Thales Electron Devices [BL07] [Clo01]. The relation between absorbed power and conductivity is defined in equation A.1. This factor is obtained from parabolic curve fitting of the experimental results up to 600 °C with the reference of copper at 0 °C. $P_{a,cu}(0\ °C)$ and $P_{a,Glidcop}$ represent the total wall loading for ideal copper at 0 °C and Glidcop material. The relation between Glidcop conductivity and temperature can be obtained as equation A.2.

$$\frac{P_{a,\,cu}(0\ °C)}{P_{a,Glidcop}} = \sqrt{\frac{\sigma_{Glidcop}}{\sigma_{cu}(0\ °C)}} = a_0 - a_1 \cdot T + a_2 \cdot T^2 < 1 \qquad (A.1)$$

$$f_G(T) = \sigma_{Glidcop} = \sigma_{cu}(0\ °C) \cdot \left(a_0 - a_1 \cdot T + a_2 \cdot T^2\right)^2 \qquad (A.2)$$

As the cavity wall thickness is many times larger than the skin depth, the effect of surface roughness can be calculated using the Hammerstad-and-Bekkadal formula [HB75] [TBDG10], as present in equation A.3. Here, K_{sr} is the enhancement factor which is the ratio of power loss at a rough surface to the power loss at the corresponding smooth surface, h is the rms height of the rough surface profile. δ is the skin depth, which depends on the angular frequency ω_{RF} of the incident wave and the magnetic permeability (μ) of the material (see equation A.4).

$$K_{sr} \equiv \frac{P_{\text{loss,rough}}}{P_{\text{loss,smooth}}} = 1 + \frac{2}{\pi}\arctan\left(\sqrt{2}\left(\frac{h}{\delta}\right)^2\right) \tag{A.3}$$

$$\delta = \sqrt{\frac{2}{\sigma\ \omega_{\text{RF}}\ \mu}} \tag{A.4}$$

The combined formula to calculate the effective conductivity of Glidcop considering the operating frequency, temperature and surface roughness dependency is,

$$\sigma_{\text{Glidcop,eff}} = \frac{f_G(T)}{\left[1 + \frac{2}{\pi} \cdot \arctan\left\{\frac{\omega_{\text{RF}}\ \mu h^2}{\sqrt{2}} \cdot f_G(T)\right\}\right]^2}. \tag{A.5}$$

In Figure A.1, the temperature dependency of the Glidcop conductivity is presented for various cases. The typical value of surface roughness (h = 0.1 μm) have been used for this analysis. The conductivity is highly de-

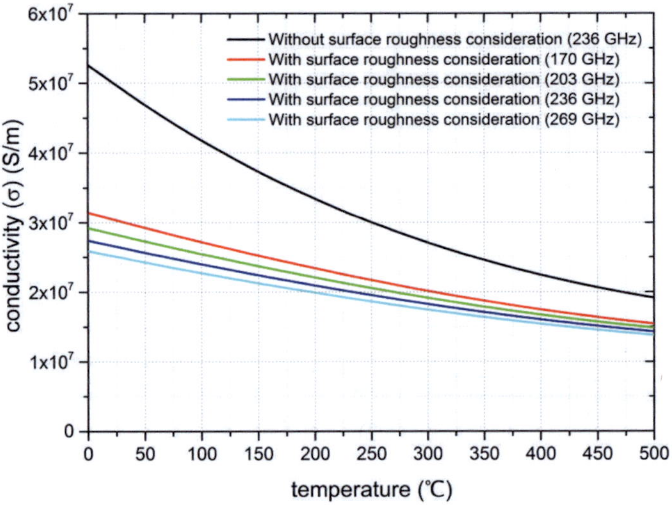

Figure A.1: Effects of surface roughness on Glidcop conductivity.

pendent on the surface roughness, while weakly dependent on the operating frequency and temperature. This method is effective to estimate realistic Glidcop conductivity at particular operating temperature, frequency and surface roughness. By suitable selection of the material specific temperature dependent function ($f_G(T)$), the obtained formula (equation A.5) could also be valid for the other cavity-materials.

A.2 Features of EURIDICE code-package

The EURIDICE code is suitable to estimate an RF-behavior of a metallic cylindrically symmetric resonator for hollow and coaxial cavity gyrotrons with finite conductivity. In the latest version of this code, it is also possible to analyze deformed cylindrical structures, which is helpful to investigate the cavity performance and frequency shift during the long-pulse operation. A frequency and surface-roughness dependent conductivity can also be calculated.

As the code only deals with classical TE modes without including the possibilities of mode conversion, the axial non-uniformity of less than 5° is preferred to avoid mode conversion. The effects of the corrugated co-axial insert have been incorporated using the Surface Impedance Model (SIM). Along with the interaction at fundamental mode, it is also possible to study mode interactions at higher harmonics.

Along the resonator axis, constant or realistic magnetic-field profile can be selected for the simulation. The implementation of After Cavity Interaction (ACI) is adopted from [Cho14]. An electron beam is represented with a discrete number of beamlets using macro-electrons (refer Figure 4.1), which can be either generated internally or the beam properties can be imported from an external input file. The realistic beam properties, i.e. energy spread, velocity spread and radial width etc. shall also be introduced in concentric electron beam, but it is not possible to generate and use misaligned electron beam with realistic parameters.

The MATLAB code-package has been developed to generate a misaligned electron beam with realistic parameters and are imported in EURIDICE to investigate beam misalignment tolerances. The detailed description of each module of the EURIDICE code-package and corresponding numerical implementation is presented in [Avr06], [APIV12] and [Avr15].

A.3 Realization of velocity spread and radial width

A.3.1 Method 1: Equal area approach

The various experimental results [PIT16] [GGK$^+$99] [Xiz84] and numerical analysis [DK99] [ZZM06] [XCYH14] suggest Gaussian like distribution of gyrotron velocity spread. The equivalent velocity ratio distribution can be expressed as,

$$f(\alpha_\mathrm{i}) = \frac{1}{\sqrt{2\cdot\pi\cdot\sigma}}\left[e^{-\frac{(\alpha_\mathrm{i}-\alpha_0)^2}{2\cdot\sigma^2}}\right] \tag{A.6}$$

Where, $\sigma = \delta\alpha\cdot\alpha_0$ is a standard deviation of the Gaussian curve, which depends on the spread. As a next task, few values of the velocity ratio are sampled according to the Gaussian distribution and are assigned to the macro electrons.

In this method, according to the desired velocity spread value, a few samples of the velocity ratio (α_k) have been selected considering the Gaussian probability distribution. These are later assigned to the micro-electrons to generate a realistic electron beam. The samples of velocity ratio (α_k) are selected, which are divided by equal area under the Gaussian probability density function. As the area under probability density function (PDF) between two alpha samples represents the probability, in other words, the alpha samples are selected, which are separated by equal probability. The total area under the Gaussian probability distribution function is calculated and according to the total number of velocity ratio samples (*N*), the area between two successive samples (*A*) are calculated with the help of equation A.7. Subsequently, from the highly dense uniformly distributed points (α_i), the velocity ratio samples (α_k) are selected, which are separated by equal area under the PDF curve.

$$\frac{1}{N+1} \cdot \int_{\alpha_0 - 10\sigma}^{\alpha_0 + 10\sigma} f(\alpha_i) \cdot d\alpha_i = A = \int_{\alpha_k}^{\alpha_{k+1}} f(\alpha_i) \cdot d\alpha_i \qquad (A.7)$$

The trapezoidal integration method is used to perform discrete integration in MATLAB. Only for the better illustration of this method, the velocity spread implantation using this method is demonstrated in Figure A.2 with only 15 sampling points within the range of $\pm 3\sigma$ around the mean value of velocity ratio of 1.25. The considered transverse rms velocity spread of the Gaussian distribution is 6 %, which corresponds to the alpha spread of 15.37 %.

Using this method, the final implemented alpha spread can be determined using equation A.8 and further translated into equivalent transverse velocity spread using equation 1.12. The desired velocity spread and the implemented spread by this method is compared in Table A.1. For this analysis, 219 samples are considered out of 0.1 million points within $\pm 10\sigma$ space. The implemented velocity spread is very close to the desired spread value. The small difference in the implemented velocity spread is due to the dis-

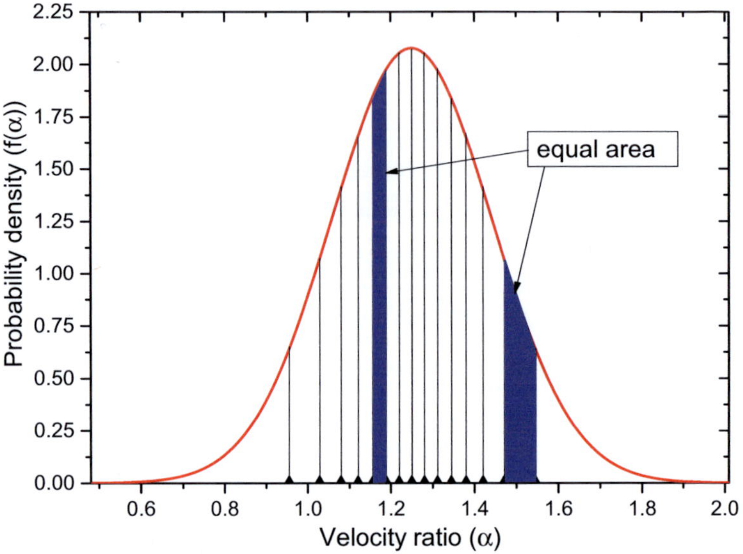

Figure A.2: Selection of velocity ratio based on equal area approach (method 1).

Table A.1: Performance analysis of the method 1. In total 219 samples are considered for the analysis. All values in the table represent the percentage rms spread.

Desired perpendicular velocity spread ($\delta\beta_{\perp}$)	Equivalent spread in velocity ratio ($\delta\alpha$)	Implemented spread in velocity ratio ($\delta\alpha_{\text{eff}}$)	Implemented spread in perpendicular velocity spread ($\delta\beta_{\perp\text{eff}}$)
2	05.125	05.084	1.98
4	10.250	10.160	3.97
6	15.375	15.320	5.97
8	20.500	20.330	7.93
10	25.625	25.534	9.96

cretization error, which will be further reduced by increasing total points α_i (> 0.1 million) and sampling points (> 219).

$$\delta\alpha_{\text{eff}} = \frac{\left(\sqrt{\frac{1}{N}\cdot\sum_{k=1}^{N}\left(\alpha_k-\alpha_0\right)^2}\right)}{\alpha_0} \tag{A.8}$$

The effect of number of samples on the spread implementation is presented in Figure A.3. The simulation setup to the calculate output power is presented in section 2.3.2. The precision of the spread implementation is highly dependent on the number of samples (α_k). For the realistic electron beam for gyrotron simulations, the more than 50 samples are required to maintain sufficient accuracy. The odd numbers of samples are preferred to keep uniform distribution from both sides of the mean value.

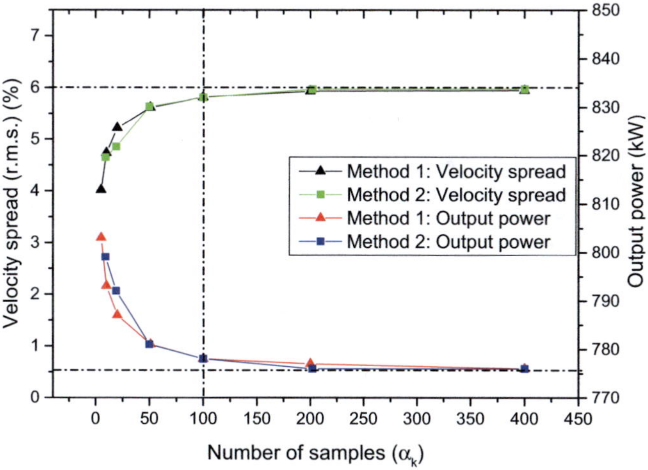

Figure A.3: Effect of number of samples on the implemented spread. For the desired 6 % of velocity spread, the performance of the method 1 and 2 is highly dependent on the number of samples considered for the analysis. More than 50 samples (α_k) are suggested to get good accuracy.

A.3.2 Method 2: Weighting factor assignment

In this method, the velocity ratio samples (α_k) are chosen uniformly within the total space of $\pm 3\sigma$ around the mean value and these selected samples are replicated according to the weighting factor of each sample. The weighting factor (W_k) of each sample is calculated considering the Gaussian distribution function. As a result, the final Gaussian distribution of the total velocity samples (N) is achieved. As described in (A.9), the normalized Gaussian function is used for this analysis.

$$f'(\alpha_i) = \left[e^{-\frac{(\alpha_i - \alpha_0)^2}{2\cdot\sigma^2}} \right] \tag{A.9}$$

The weighting factor of each sample can be calculated using equation A.10, where the value of k varies from 1 to M (number of uniform samples). The repetition number N_k of a particular uniform sample is the product of the weighting factor of each samples α_k and the total velocity samples (N). Further, the repetition number (N_k) for a particular sample (α_k) is rounded to the nearest integer.

$$W_k = \frac{f'(\alpha_k)}{\sum_{k=1}^{M} f'(\alpha_k)} \tag{A.10}$$

$$N_k = W_k \cdot N \tag{A.11}$$

$$N = \sum_{k=1}^{M} W_k \cdot N = \sum_{k=1}^{M} N_k \tag{A.12}$$

For illustration purpose, the example of this method is presented in Figure A.4 with 6 % of the perpendicular velocity spread. In total 15 uniform velocity ratio samples (α_k) are selected within the range of $\pm 3\sigma$. Based on the weighting factors (W_k), each uniform velocity sample (α_k) is repeated

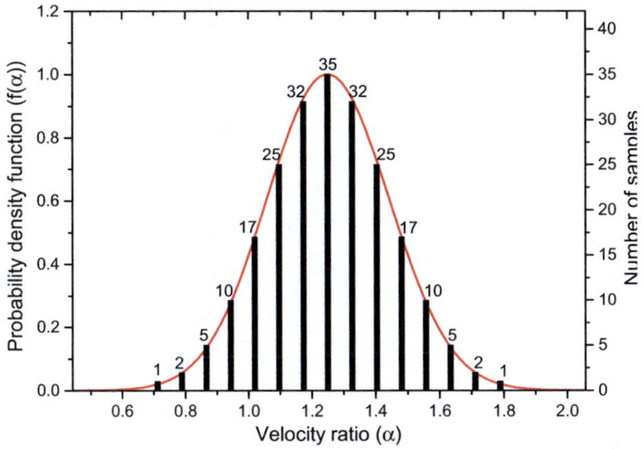

Figure A.4: Selection of the velocity ratio based on Method 2. In this case, total 219 velocity ratio samples are distributed among the 15 uniformly spaced alpha samples to achieve 6 % of perpendicular velocity spread.

N_k times to form the electron beam with total 219 samples, which are distributed according to Gaussian function.

The implemented spread using this method can be calculated using equation A.13. In Table A.2, for different perpendicular velocity spread values, the effective implemented spread with this method is listed for 21 uniform alpha samples (α_k) and 219, total number of samples. The distribution pattern perpendicular velocity spread using method 1 and method 2 are presented in Figure A.5.

$$\delta\alpha_\mathrm{eff} = \frac{\left(\sqrt{\frac{1}{N}\cdot\sum_{k=1}^{M} N_\mathrm{k}\cdot(\alpha_\mathrm{k}-\alpha_0)^2}\right)}{\alpha_0} \tag{A.13}$$

Table A.2: Performance analysis of the method 2. In total 19 samples are selected with 219 total velocity samples (N). All values in the table represent the percentage rms spread.

Desired perpendicular velocity spread ($\delta\beta_\perp$)	Equivalent spread in velocity ratio ($\delta\alpha$)	Implemented spread in velocity ratio ($\delta\alpha_{\mathrm{eff}}$)	Implemented spread in perpendicular velocity spread ($\delta\beta_{\perp\mathrm{eff}}$)
2	05.125	5.05	1.97
4	10.250	10.10	3.94
6	15.375	15.18	5.92
8	20.500	20.20	7.88
10	25.625	25.18	9.83

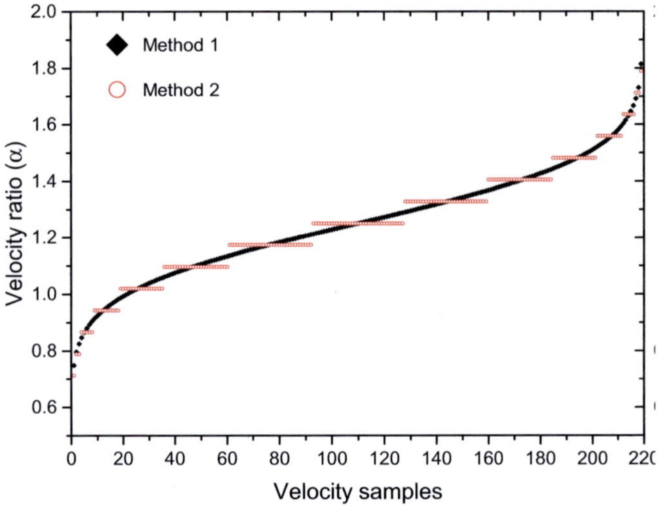

Figure A.5: Distribution pattern of the velocity ratio for both methods. In total 219 velocity ratio samples are distributed using method 1 and method 2 to achieve 15.37 % Gaussian spread.

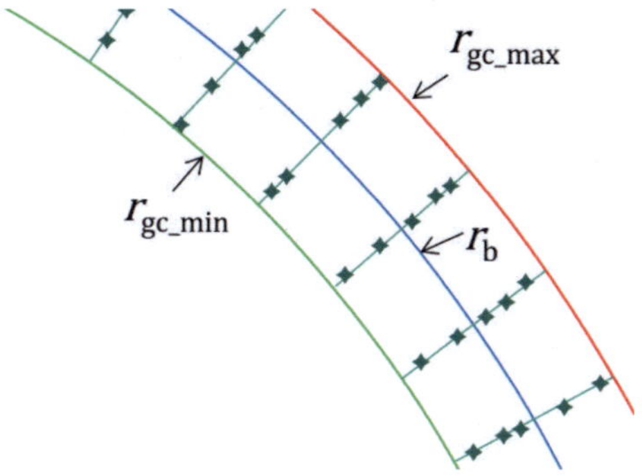

Figure A.6: Distribution of an electron guiding centers to achieve realistic beam properties with $\lambda/4$ radial width.

A.3.3 Radial width implementation

A uniform electron guiding center spread (Δr_{gc}) is introduced to implement the radial width. The maximum and minimum electron guiding center radii can be calculated by, $r_{\mathrm{gc\_max}} = r_{\mathrm{b}} + \Delta r_{\mathrm{gc}}/2$ and $r_{\mathrm{gc\_min}} = r_{\mathrm{b}} - \Delta r_{\mathrm{gc}}/2$, respectively.

In this method, all electron guiding centers are distributed randomly within these limits. The total radial thickness of the beam is described as $\Delta r_{\mathrm{b}} = \Delta r_{\mathrm{gc}} + 2r_{\mathrm{L}}$. Here, r_{L} is the Larmor radius. The probability density function of the guiding center distribution is described as

$$f\left(r_{\mathrm{gc}}\right) = \begin{cases} 1/\Delta r_{\mathrm{gc}} \ \mathrm{for}\, \mathrm{r_{gc\_min}} \leq \mathrm{r_{gc}} \leq \mathrm{r_{gc\_max}} \\ 0 \ \ \mathrm{for}\, \mathrm{r_{gc}} < \mathrm{r_{gc\_min}} \,\mathrm{and}\, \mathrm{r_{gc}} > \mathrm{r_{gc\_max}} \end{cases} \tag{A.14}$$

The implemented distribution of the guiding centers is presented in Figure A.6 with distribution of 5 guiding centers in each azimuthal direction. The electron guiding centers are distributed randomly between the minimum guiding center radius $r_{\mathrm{gc\_min}}$ and $r_{\mathrm{gc\_max}}$. For the different values, the comparison of the desired radial width and the implemented radial width

Table A.3: Performance analysis of the radial width implementation. Azimuthal space is divided into 50 equal parts and 20 guiding centers are distributed per angle.

Desired radial width	Implemented radial width
$\lambda/16$	$\lambda/16.097$
$\lambda/8$	$\lambda/8.009$
$\lambda/4$	$\lambda/4.0025$
$\lambda/2$	$\lambda/2.006$
λ	$\lambda/1.0011$

is presented in Table A.3. The good agreement between both the values supports the suggested method for radial width implementation.

A.3.4 Validation of the velocity spread and radial width implementation

Comparison of velocity spread implementation

For 236 GHz DEMO gyrotron, the single mode time-dependent simulations are used to validate the velocity spread implementation methods discussed in section A.3.1 and A.3.2. The spread implementation using macro-electron beam are compared with the EURIDICE internal spread implementation approach. During the start-up scenario, the beam energy is increased from 28 keV to 58 keV and the RF behavior of gyrotron is estimated. The selected operating parameters for this analysis are enlisted in section 2.3.2. In all the cases, 6 % (r.m.s.) perpendicular velocity spread $\beta_\perp$ is considered, which corresponds to 15.37 % of spread in the velocity ratio (α_0). The comparison of final results is presented in Figure A.7.

For the equal area method (method 1), 219 samples are considered to implement the Gaussian distribution, and as discussed in section 4.1.2, the

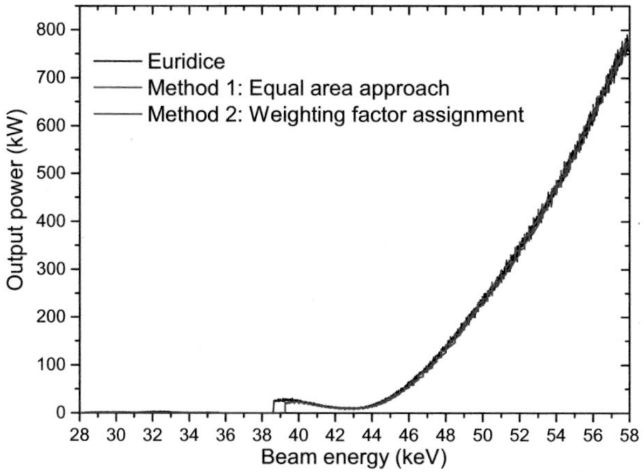

Figure A.7: Single mode, time-dependent simulations to validate developed spread implementation algorithms. Using method 1 and method 2, the macro-electron beam with 6 % (rms), Gaussian velocity spread is generated and used for the simulations.

macro-electron beam is generated with 50 guiding center phases ϕ_{gc} and 9 different phases (ϕ). The macro-electron beam with 6 % velocity spread is further used for single mode simulation in EURIDICE code. The radial width is not considered in this simulation. Similarly, using weighting factor assignment, 219 samples are distributed to 21 uniform velocity ratio sample (α_k). The results with suggested velocity spread implementation methods are in good agreement with EURIDICE spread implementation and authenticates the suggested approaches. Along with the single mode simulations, the performance of suggested methods have also been successfully verified/benchmarked with multi-mode simulations.

For further verification, the velocity spread implementation of the EURIDICE code package is compared with TWANG and TWANG-PIC codes. Using the test case of W7-X gyrotron and the ITER gyrotron, the effects of beam velocity spared are investigate for all three codes. The detailed set-up of these cases and results are discussed in section A.5.

Radial width implementation

Similar to the velocity spread, the radial width implementation using macro-electron beam (section A.3.3) is compared to the standard EURIDICE internal width implementation. In Figure A.8, the single mode self-consistent simulation result of the radial width implementation with macro-electron beam is benchmarked with the EURIDICE radial width implementation. The beam width of $\lambda/4$ is considered for both the simulations. For the particular guiding center phase (ϕ_{gc}), 20 different guiding center radii (r_{gc}) are used to generate realistic electron beam. The effects of radial width are identical in both the cases and support the suggested radial width implementation method discussed in section A.3.3.

A.4 Developed code package: VISTAR

Using discussed techniques of the velocity spread and the radial width implementation, the MATLAB code package named "VISTAR" has been developed. Using this code package, the realistic electron beam file has

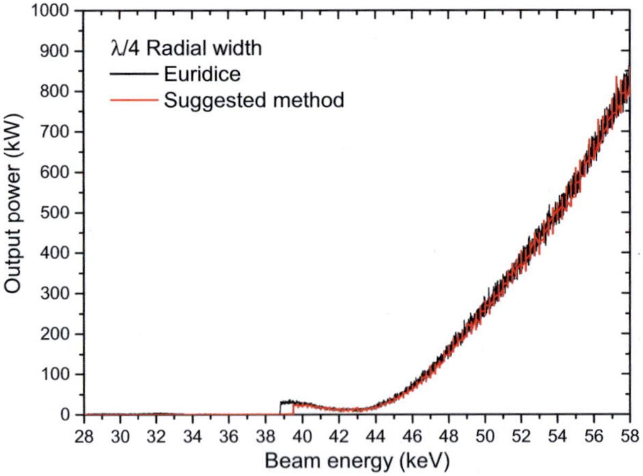

Figure A.8: Single-mode, time-dependent simulations with a beam thickness of $\lambda/4$.

been generated and further used as an input for the self-consistent, time-dependent simulations or PIC simulations. The final realistic electron beam file is fully compatible with in-house beam-wave interaction simulation code "EURIDICE" [APIV12] along with the commercially available gyrotron interaction codes (e.g. MAGIC [All11], CST PARTICLE STUDIO® [CST] etc.). By generating the misaligned realistic electron beam file, the effects of axial beam misalignment can also be studied. In addition to that, it is also possible to customize the implemented distribution function (i.e. Gaussian for velocity spread and uniform for radial width) and the new user defined distribution function can be easily replaced.

Major functions of VISTAR code

The VISTAR code includes many individual sub-routings. The important sub-routings and their functionalities are as discussed below.

- **Calculate the beam parameter for the start-up condition:** Using the final operating parameters (beam voltage, beam current and velocity ratio), this code calculates the beam parameters during the start-up scenario.
- **Generate electron beam file:** Considering the electron beam radius, parameters of macro-electron beam and start-up scenario, an ideal macro-electron beam file can be generated.
- **Analyze beam parameters:** It is possible analyze the electron beam file and identify the various beam properties such as: beam position, a variation of the relativistic factor (γ), various spreads in the beam parameters etc..
- **Introduce velocity spread:** As discussed in section A.3.1 and A.3.2, two methods have been selected to generate electron beam with velocity spread. Both methods are implemented successfully in VISTAR

code and an electron beam with velocity spread could be generated using the developed codes.

- **Radial width implementation:** The codes for radial width implementation have been successfully developed and integrated in VISTAR. The numerical analysis of the implemented method is discussed in section A.3.3.
- **Introduce beam misalignment:** This part of code generates misaligned ideal electron beam (without velocity spread and radial width) for the electron beam misalignment study. As a next step, the velocity spread and radial width implementation routings are used to generate a misaligned electron beam with realistic beam parameters for the accurate estimate of gyrotron behavior. The detailed analysis of the electron beam misalignment is presented in section 4.1.
- **Export file with the realistic beam profile:** The generated realistic electron beam file is further exported for gyrotron interaction simulations. The generated file is compatible with the self-consistent EURIDICE code as well as with the PIC codes like MAGIC [All11] and CST PARTICLE STUDIO® [CST].

A.5 Comparison of velocity spread implementation: EURIDICE and TWANG code

For the accurate estimation of the gyrotron behavior, it is necessary to consider the effects of electron velocity spread. Over the period, various gyrotron interaction codes have been developed at IHM-KIT Germany, National Technical University of Athens, Greece and SPC-EPFL, Switzerland. The main aim of this analysis is to compare/benchmark the velocity spread implementation of these in-house codes. For comparison, the cases of the

W7-X 140 GHz / 1 MW gyrotron and the ITER 170 GHz / 1 MW gyrotron are chosen. The complete parameters of the selected cases are enlisted in the Table A.4. The co-rotating mode $TE_{-28,8}$ is selected for W7-X case and co-rotating $TE_{-32,9}$ mode for the ITER case. In case of the W7-X gyrotron, the constant magnetic of 5.59 T is considered, while in the case of ITER gyrotron, the realistic magnetic field profile is used.
The Gaussian-type distribution is selected for the pitch-factor spread. The simulations are performed with the pitch-factor spread of 5 %, 10 %, 15 % and 20 %, which corresponds to the perpendicular velocity spread of 1.85 %, 3.72 %, 5.57 % and 7.43 %, respectively. Here, the average value of pitch factor α_0 is 1.3.

Simulations with Cavity code and EURIDICE

To validate spread implementation, initially, the single mode self-consistent, time-dependent simulations are performed using in-house codes, Cavity and EURIDICE. In both the cases (W7-X and ITER), the gyrotron performance

Table A.4: Selected modes and the operating parameters for the W7-X and ITER test case.

	Case 1: W7-X	Case 2: ITER
Frequency	140 GHz	170 GHz
Selected operating mode	$TE_{-28,8}$	$TE_{-32,9}$
Magnetic field (T)	5.59	6.78
Cavity radius (mm)	20.48	19.24
Beam radius (mm)	10.1	9.44
Beam voltage (keV)	80	79.5
Beam current (A)	40	40
Pitch factor	1.3	1.3
Conductivity (S/m)	1.4×10^7	1.4×10^7

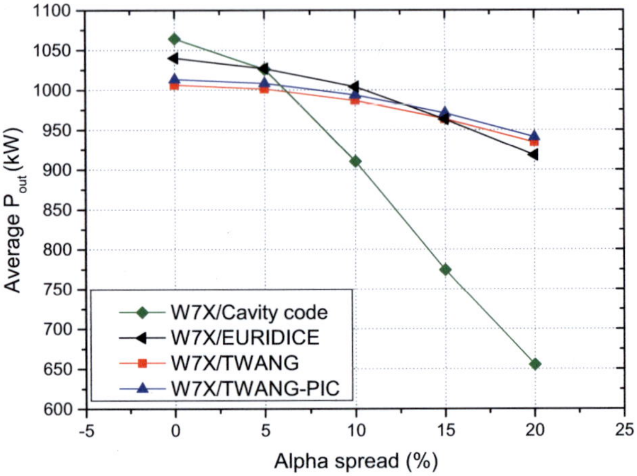

(a)

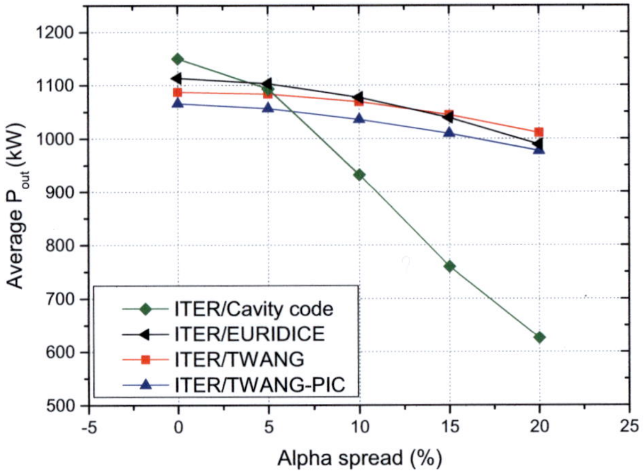

(b)

Figure A.9: RF-power versus velocity-ratio spread on the test case of the (a) W7-X and (b) ITER gyrotron.

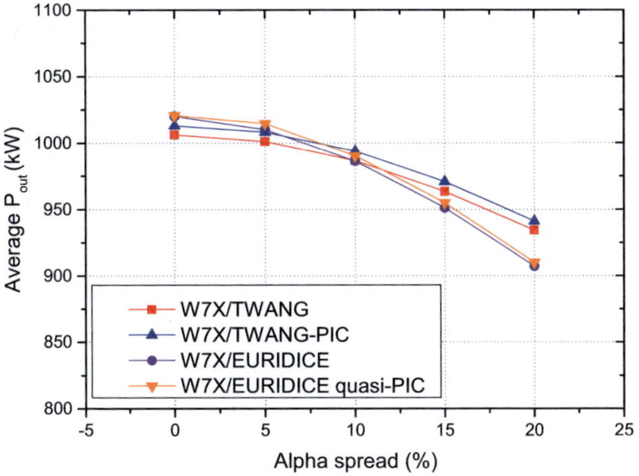

(a)

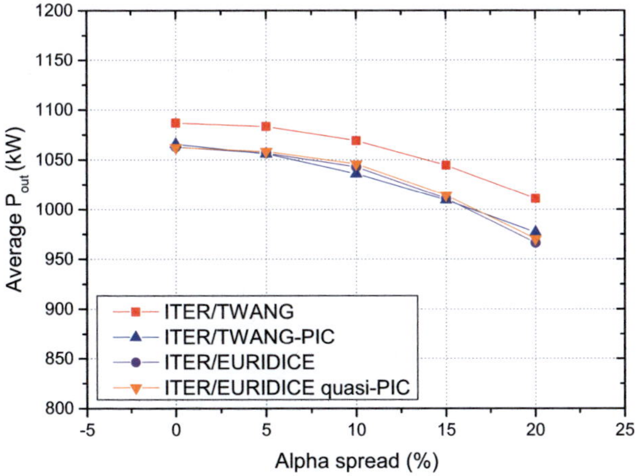

(b)

Figure A.10: Dependency of the output power on the pitch factor spread in different codes considering the variation of axial velocity along the gyrotron The simulation results of (a) W7-X and (b) ITER case are compared.

is determined by considering different velocity spread values. The dependency of the output power of W7-X gyrotron on the pitch factor spread is presented in Figure A.9 (a). The output power is highly dependent in the case of Cavity code simulations, and there is a discrepancy in the results of the Cavity code and EURIDICE. The results are similar to the ITER case also (see Figure A.9 (b)). These scenarios strongly suggest the necessity of further efforts to validate the spread implementation.

Benchmark results with TWANG and TWANG-PIC code

To resolve the discrepancies between the results of the Cavity code and EURIDICE, the influence of velocity spread is verified with the simulation results of TWANG and TWANG-PIC code [ATA^{+}11] [BTV^{+}14] [BTV^{+}15]. The same W7-X and ITER cases are used for the simulations. The simulations with TWANG and TWANG-PIC code have been performed by Dr. Falk Braunmüller [Bra15].

For all the selected codes, the output power at different spread values is compared in Figure A.9. The power without any spread is slightly smaller in the TWANG simulations than from the KIT-codes simulations. The two codes TWANG and TWANG-PIC have practically the same dependency on spread, which is even smaller than the one from EURIDICE. Clearly, the results from Cavity code are again outlier. TWANG-and TWANG-PIC seem to have exactly the same dependency on the pitch-angle.

It is observed that the dependency on the pitch factor spread is more for SPC-codes TWANG and TWANG-PIC rather than EURIDICE. However, the dependency of these three codes are in the same order and the the results form Cavity-codes are much more dependent on the spreads.

Effects of axial velocity variation

As presented in Figure A.9, the dependency of the pitch-angle spread is similar in EURIDICE, TWANG and TWANG-PIC simulations, though, the output power in the case of EURIDICE marginally differs to TWANG and TWANG-PIC simulations. The axial velocity of the electrons along with the cavity has been considered constant in EURIDICE simulations, while the variation of axial velocity is included in TWANG and TWANG-PIC simulations. To have similar simulation setup, the EURIDICE simulations are performed again including the effects of axial velocity variation. In addition to that, the similar simulations are also performed using EURIDICE quasi-PIC routing for better comparison. The dependence of the pitch - factor spread of the output power with new updated EURIDICE and EURIDICE quasi-PIC is compared with TWANG and TWANG-PIC codes as presented in Figure A.10.

With the inclusion of axial velocity variation in EURIDICE simulations, the overall output power is slightly reduced in both the cases (W7-X and ITER). The results of EURIDICE and EURIDICE quasi-PIC routine are in good agreement for different values of pitch-factor spread. For the ITER gyrotron, with the updated simulation setup, the results of the EURIDICE and EURIDICE quasi-PIC simulations are similar to TWANG code.

Bibliography

[ADV08] K. A. Avramides, O. Dumbrajs, and J. L. Vomvoridis. Mode selection for a 170 GHz, 1 MW gyrotron. *35th EPS Conference on Plasma Phys.*, 2008.

[AEZS77] I. I. Antakov, V. S. Ergakov, E. V. Zasypkin, and E. V. Sokolov. Starting conditions of a CRM monotron in the presence of scatter of the velocities of the electrons. *Radiophysics and Quantum Electronics*, 20(4):413–418, 1977.

[AGM^{+}12] G. Aiello, G. Grossetti, A. Meier, T. Scherer, S. Schreck, P. Spaeh, D. Strauss, and A. Vaccaro. CVD diamond brewster window: feasibility study by FEM analyses. *EPJ Web of Conferences*, 32:04014, 2012.

[Air03] M. I. Airila. Degradation of operation mode purity in a gyrotron with an off-axis electron beam. *Physics of Plasmas*, 10(1):296–299, 2003.

[All11] Alliant Techsystems (ATK), Virginia. *MAGIC 3.2.0 Help MANUAL*, October 2011.

[APIT13] K. A. Avramidis, I. Gr. Pagonakis, Z. C. Ioannidis, and I. G. Tigelis. Numerical investigations on the effects of electron beam misalignment on beam-wave interaction in a high-power coaxial gyrotron. In *38th International Conference on Infrared, Millimeter, and Terahertz Waves (IRMMW-THz)*, 2013.

[APIV12] K. A. Avramides, I. Gr. Pagonakis, C. T. Iatrou, and J. L. Vomvoridis. EURIDICE: A code-package for gyrotron interaction simulations and cavity design. *EPJ Web of Conferences*, 32:04016, 2012.

[ATA+11] S. Alberti, T. M. Tran, K. A. Avramides, F. Li, and J.-P. Hogge. Gyrotron parasitic-effects studies using the time-dependent self-consistent monomode code TWANG. In *36th International Conference on Infrared, Millimeter, and Terahertz waves (IRMMW-THz)*, 2011.

[Avr06] K. A. Avramides. *Design and simulation of coaxial gyrotrons (laying emphasis on second-harmonic operation) (in Greek)*. PhD thesis, National Technical University of Athens, 2006.

[Avr15] K. A. Avramidis. Investigations and advanced concepts on gyrotron interaction modeling and simulations. *Physics of Plasmas*, 22(12):123114, 2015.

[BAL+98] M. Botton, T.M. Antonsen, B. Levush, K.T. Nguyen, and A.N. Vlasov. MAGY: a time-dependent code for simulation of slow and fast microwave sources. *IEEE Transactions on Plasma Science*, 26(3):882–892, 1998.

[BDKT97] O. Braz, G. Dammertz, M. Kuntze, and M. Thumm. D-band frequency step-tuning of a 1 MW gyrotron using a brewster output window. *International Journal of Infrared and Millimeter Waves*, 18(8):1465–1477, 1997.

[BDN16] V. L. Bakunin, G. G. Denisov, and Yu. V. Novozhilova. Zones of frequency locking by an external signal in a multimode gyrotron of a megawatt power level. *Radiophysics and Quantum Electronics*, 58(12):893–904, 2016.

[Ber10] M. Beringer. *Design Studies towards a 4 MW 170 GHz Coaxial-Cavity Gyrotron.* PhD thesis, Karlsruher Institut für Technologie (KIT), 2010.

[BG91] E. Borie and G. Gantenbein. Self consistent theory for gyrotrons including effect of voltage depression. *International Journal of Infrared and Millimeter Waves*, 12(2):65–78, 1991.

[BGG+00] R. J. Buttery, S. Günter, G. Giruzzi, T. C. Hender, D. Howell, G. Huysmans, R. J. La. Haye, M. Maraschek, H. Reimerdes, O. Sauter, C. D. Warrick, H. R. Wilson, and H. Zohm. Neoclassical tearing modes. *Plasma Physics and Controlled Fusion*, 42(12B):B61–B73, 2000.

[BJ87] E. Borie and B. Jödicke. Startup and mode competition in a 150 GHz gyrotron. *International journal of infrared and millimeter waves*, 8(3):207–226, 1987.

[BJ88] E. Borie and B. Jodicke. Comments on the linear theory of the gyrotron. *IEEE Transactions on Plasma Science*, 16(2):116–121, 1988.

[BKC99] J. J. Barroso, K.G. Kostov, and R.A. Correa. Electromagnetic simulation of a 32 GHz, TE_{021} gyrotron. *IEEE Transactions on Plasma Science*, 27(2):384–390, 1999.

[BKS16] I. V. Bandurkin, Yu. K. Kalynov, and A. V. Savilov. Experimental study of a gyrotron with a sectioned klystron-type cavity operated at higher cyclotron harmonics. *Radiophysics and Quantum Electronics*, 58(9):694–700, 2016.

[BKT94] M. Blank, K.E. Kreischer, and R.J. Temkin. Study of rotating modes in high frequency whispering gallery mode gyrotrons. *IEEE Transactions on Plasma Science*, 22(5):883–888, 1994.

[BL07] D. Bariou and C. Lievin. Design report ITER 170-GHz gyrotron contract no. efda-03/960. Technical report, Thales Electron Devices, RT5307 (Internal Report), 2007.

[Bor91] E. Borie. Review of gyrotron theory. *KfK-4898*, 1991.

[Bra83] G. F. Brand. A gyrotron frequency detuning equation. *International journal of infrared and millimeter waves*, 4(6):919–931, 1983.

[Bra15] F. Braunmueller. presentation at 4th progress meeting of. Grant agreement: F4E-GRT-553, Design and Development of the EU gryrotron, Karlsruhe, Germany., Sept. 2015.

[BSC16] V. L. Bratman, A. V. Savilov, and T. H. Chang. Possibilities for continuous frequency tuning in terahertz gyrotrons with non-tunable electrodynamic systems. *Radiophysics and Quantum Electronics*, 58(9):660–672, 2016.

[BTV^{+}14] F. Braunmueller, T. M. Tran, Q. Vuillemin, S. Alberti, J. Genoud, J. Ph. Hogge, and M. Q. Tran. TWANG-PIC, a monomode gyro-averaged PIC code for gyrotron simulations. In *39th International Conference on Infrared, Millimeter, and Terahertz waves (IRMMW-THz)*, 2014.

[BTV^{+}15] F. Braunmueller, T. M. Tran, Q. Vuillemin, S. Alberti, J. Genoud, J.-Ph. Hogge, and M. Q. Tran. TWANG-PIC, a novel gyro-averaged one-dimensional particle-in-cell code for interpretation of gyrotron experiments. *Physics of Plasmas*, 22(6):063115, 2015.

[But84] K. J. Button. *Infrared and Millimeter Waves V11: Millimeter Components and Techniques*. Elsevier, 1984.

[CBB+11] S. Cauffman, M. Blank, P. Borchard, P. Cahalan, and K. Felch. Development of a 170 GHz, 500 kW CW gyrotron for testing ITER transmission line components. In *2011 International Conference on Infrared, Millimeter, and Terahertz Waves*, 2011.

[CBF+07] S. Cauffinan, M. Blank, K. Felch, P. Borchard, P. Cahalan, and H. Jory. Initial testing of a 95 GHz, 2.5 MW gyrotron. In *Joint 32rd International Conference on Infrared and Millimeter Waves, 2007 and the 15th International Conference on Terahertz Electronics.*, pages 94–95, 2007.

[CDK+06] A. V. Chirkov, G. G. Denisov, M. L. Kulygin, V. I. Malygin, S. A. Malygin, A. B. Pavel'ev, and E. A. Soluyanova. Use of huygen's principle for analysis and synthesis of the fields in oversized waveguides. *Radiophysics and Quantum Electronics*, 49(5):344–353, 2006.

[Che74] F. F. Chen. *Introduction to plasma physics*. Plenum Press,, 1974.

[Che11] F. F. Chen. *An Indispensable Truth*. Springer New York, 2011.

[Cho14] A. R. Choudhury. *Investigations of After Cavity Interaction in Gyrotrons Including the Effect of Non-uniform Magnetic Field*. PhD thesis, Karlsruher Institute für Technologie, 2014.

[Chu04] K. R. Chu. The electron cyclotron maser. *Reviews of modern physics*, 76(2):489, 2004.

[CIO+09] T. H. Chang, T. Idehara, I. Ogawa, L. Agusu, and S. Kobayashi. Frequency tunable gyrotron using backward-wave components. *Journal of Applied Physics*, 105(6):063304, 2009.

[CJ12] B. R. Chandra and P. K. Jain. Multimode behavior of a 42 GHz, 200 kW gyrotron. *Progress in Electromagnetics Research B*, 42:75, 2012.

[Clo01] G. L. Clorarec. Characterisation des anneaux de raschig dans une structure cavite-diode validation du choix pour le gyrotron 140 GHz, 2001.

[CMM+06] E. M. Choi, C. D. Marchewka, I. Mastovsky, J. R. Sirigiri, M. A. Shapiro, and R. J. Temkin. Experimental results for a 1.5 MW, 110 GHz gyrotron oscillator with reduced mode competition. *Physics of Plasmas*, 13(2):023103, 2006.

[CS81] J. Y. Choe and A. Saeyoung. General mode analysis of a gyrotron dispersion equation. *IEEE Transactions on Electron Devices*, 28(1):94–102, 1981.

[CST] CST - Computer Simulation Technology, Darmstadt Germany. *CST STUDIO SUITE user's Manual.*

[DII+03] O. Dumbrajs, T. Idehara, Y. Iwata, S. Mitsudo, I. Ogawa, and B. Piosczyk. Hysteresis-like effects in gyrotron oscillators. *Physics of Plasmas*, 10(5):1183–1186, 2003.

[DIP+05] G. Dammertz, S. Illy, B. Piosczyk, M. Schmid, and D. Bariou. Collector sweeping systems for high power gyrotrons. In *Joint 30th International Conference on Infrared and Millimeter Waves and 13th International Conference on Terahertz Electronics*, 2005.

[DIW+04] O. Dumbrajs, T. Idehara, S. Watanabe, A. Kimura, H. Sasagawa, L. Agusu, S. Mitsudo, and B. Piosczyk. Reflections in gyrotrons with axial output. *IEEE Transactions on Plasma Science*, 32(3):899–902, 2004.

[DK81] A. T. Drobot and K. Kim. Space charge effects on the equilibrium of guided electron flow with gyromotion. *International Journal of Electronics*, 51(4):351–367, 1981.

[DK99] O. Dumbrajs and J. P. T. Koponen. Generalized gyrotron theory with inclusion of electron velocity and energy spreads. *Physics of Plasmas*, 6(6):2618–2621, 1999.

[DKM$^+$92] G. G. Denisov, A. N. Kuftin, V. I. Malygin, N. P. Venediktov, D. V. Vinogradov, and V. E. Zapevalov. 110 GHz gyrotron with a built-in high-efficiency converter. *International Journal of Electronics*, 72(5-6):1079–1091, 1992.

[DLA$^+$08] G.G. Denisov, A.G. Litvak, M.V. Agapova, V.E. Myasnikov, E.M. Tai, V.E. Zapevalov, A.V. Chirkov, A.N. Kuftin, S.A. Malygin, V.I. Malygin, V.O. Nicniporenko, I.V. Kazansky, A.V. Kruglov, V.G. Rukavishnikova, A.F. Gnedenkov, A.B. Pavel'ev, V.V. Parshin, L.G. Popov, E.V. Sokolov, E.A. Soluyanova, V.I. Ilin, V.N. Ilin, A.L. Vikharev, N.A. Shamanova, and S.V. Usachev. Multi-frequency gyrotrons for plasma fusion installations. In *33rd International Conference on Infrared, Millimeter and Terahertz Waves (IRMMW-THz)*, 2008.

[DN10] O. Dumbrajs and G. S. Nusinovich. To the theory of high-power gyrotrons with uptapered resonators. *Physics of Plasmas*, 17(5):053104, 2010.

[DS92] O. Dumbrajs and L. Shenggang. Kinetic theory of electron-cyclotron resonance masers with asymmetry of the electron beam in a cavity. *IEEE Transactions on Plasma Science*, 20(3):126–132, 1992.

[DT86] B. G. Danly and R. J. Temkin. Generalized nonlinear harmonic gyrotron theory. *Physics of Fluids*, 29(2):561, 1986.

[Dum94] O. Dumbrajs. Eccentricity of the electron beam in a gyrotron cavity. *International Journal of Infrared and Millimeter Waves*, 15(7):1255–1262, 1994.

[EBG+14] V. Erckmann, H. Braune, G. Gantenbein, J. Jelonnek, W. Kasparek, H. P. Laqua, C. Lechte, N. B. Marushchenko, G. Michel, B. Plaum, M. Thumm, M. Weissgerber, R. Wolf, and the W7-X ECRH-teams at IPP Greifswald. ECRH and W7-X: An intriguing pair. In *AIP Conf. Proc. 1580, 542 (2014)*, 2014.

[EBL+05] V. Erckmann, H. Braune, H. P. Laqua, G. Michel, M. Weissgerber, P. Brand, G. Dammertz, G. Gantenbein, W. Kasparek, and M. Thumm. The 140 GHz, 10 MW, CW ECRH system for W7-X: a training field for iter. In *The Joint 30th International Conference on Infrared and Millimeter Waves and 13th International Conference on Terahertz Electronics.*, volume 1, pages 1–2, 2005.

[Edg80] C. J. Edgcombe. The dispersion equation for the gyrotron amplifier. *International Journal of Electronics*, 48(6):471–486, 1980.

[Edg93] C. Edgecombe. *Gyrotron Oscillators*. CRC Press, 1993.

[Ein35] A. Einstein. Elementary derivation of the equivalence of mass and energy. *Bull. Amer. Math. Soc.*, 41(4):223–230, 1935.

[EKP+12] V. Erckmann, W. Kasparek, B. Plaum, C. Lechte, M. I. Petelin, H. Braune, G. Gantenbein, H. P. Laqua, L. Lubiako, N. B. Marushchenko, et al. Large scale CW ECRH systems: some considerations. In *EPJ Web of Conferences*, volume 32, 2012.

[ESL+07] V. Erckmann, M. Schmid, H. P. Laqua, G. Dammertz, S. Illy, H. Braune, and F. Hollmann. Advanced gyrotron collector sweeping with smooth power distribution. In *4th IAEA Technical Meeting on ECRH Physics and Technology for ITER*, 2007.

[FAG+14] J. Franck, K. Avramidis, G. Gantenbein, S. Illy, J. Jin, M. Thumm, and J. Jelonnek. A generic mode selection strategy for high-order mode gyrotrons operating at multiple frequencies. *Nuclear Fusion*, 55(1):013005, 2014.

[FAG+16] J. Franck, K. A. Avramidis, G. Gantenbein, S. Illy, I. Gr. Pagonakis, M. Thumm, and J. Jelonnek. Direct voltage depression calculation of arbitrary electron beams in misaligned coaxial gyrotron cavities. *IEEE Transactions on Electron Devices*, 63(9):3740–3746, 2016.

[FAP+15] J. Franck, K. A. Avramidis, I. Gr. Pagonakis, S. Illy, G. Gantenbein, M. Thumm, and J. Jelonnek. Multi-frequency design of a 2 MW coaxial-cavity gyrotron for DEMO. In *40th International Conference on Infrared, Millimeter, and Terahertz waves (IRMMW-THz)*, 2015.

[FBB+16] G. Federici, C. Bachmann, W. Biel, L. Boccaccini, F. Cismondi, S. Ciattaglia, M. Coleman, C. Day, E. Diegele, T. Franke, M. Grattarola, H. Hurzlmeier, A. Ibarra, A. Loving, F. Maviglia, B. Meszaros, C. Morlock, M. Rieth, M. Shannon, N. Taylor, M.Q. Tran, J.H. You, R. Wenninger, and L. Zani. Overview of the design approach and prioritization of R&D activities towards an EU DEMO. *Fusion Engineering and Design*, 109-111:1464–1474, 2016.

[FDJ+99] K. L. Felch, B. G. Danly, H. R. Jory, K. E. Kreischer, W. Lawson, B. Levush, and R. J. Temkin. Characteristics and applications of fast-wave gyrodevices. *Proceedings of the IEEE*, 87(5):752–781, 1999.

[FGL$^+$13] G. Federici, G. Giruzzi, C. Lowry, R. Kemp, D. Ward, R. Wenninger, and H. Zohm. EU DEMO design and R&D studies. In *25th Symposium on Fusion Engineering (SOFE)*, 2013.

[FGPY77] V. A. Flyagin, A. V. Gaponov, I. Petelin, and V. K. Yulpatov. The gyrotron. *IEEE Transactions on Microwave Theory and Techniques*, 25(6):514–521, 1977.

[FIK$^+$14] J. Franck, S. Illy, Avramidis K., J. Jelonnek, and M. Thumm. Mode selection and resonator design for demo gyrotrons. In *15th International Vacuum Electronics Conference (IVEC)*, pages 31–32, 2014.

[FKW$^+$14] G. Federici, R. Kemp, D. Ward, C. Bachmann, T. Franke, S. Gonzalez, C. Lowry, M. Gadomska, J. Harman, B. Meszaros, C. Morlock, F. Romanelli, and R. Wenninger. Overview of EU DEMO design and R&D activities. *Fusion Engineering and Design*, 89(7-8):882–889, 2014.

[FRC$^+$12] F. Felici, J.X. Rossel, G. Canal, S. Coda, B.P. Duval, T.P. Goodman, Y. Martin, J-M. Moret, O. Sauter, and D. Testa and. Real-time control of multiple MHD instabilities on TCV by ECRH/ECCD. *EPJ Web of Conferences*, 32:02005, 2012.

[Fus12] *Fusion Electricity, A roadmap to the realisation of fusion energy*. European Fusion Development Agreement (EFDA), 2012.

[FZZ$^+$09] K.M. Feng, G.S. Zhang, G.Y. Zheng, Z. Zhao, T. Yuan, Z.Q. Li, G.Z. Sheng, and C.H. Pan. Conceptual design study of fusion DEMO plant at SWIP. *Fusion Engineering and Design*, 84(12):2109–2113, 2009.

[GBB+15] S. Garavaglia, W. Bin, A. Bruschi, G. Granucci, G. Grossetti, J. Jelonnek, A. Moro, N. Rispoli, D. Strauss, Q. M. Tran, and T. Franke. Preliminary conceptual design of DEMO EC system. In *AIP Conference Proceedings 1689, 090009*. AIP Publishing, 2015.

[GDJ+13] G. Gantenbein, G. Dammertz, J. Jelonnek, M. Losert, A. Samartsev, A. Schlaich, T. Scherer, D. Strauss, M. Thumm, and D. Wagner. Operation of a step-frequency tunable gyrotron with a diamond brewster angle output window. In *2013 IEEE 14th International Vacuum Electronics Conference (IVEC)*, 2013.

[GGK+99] M.Yu. Glyavin, A.L. Goldenberg, A.N. Kuftin, V.K. Lygin, A.S. Postnikova, and V.E. Zapevalov. Experimental studies of gyrotron electron beam systems. *IEEE Transactions on Plasma Science*, 27(2):474–483, 1999.

[GKVZ97] M. Yu. Glyavin, A. N. Kuftin, N. P. Venediktov, and V. E. Zapevalov. Experimental investigation of a 110 GHz/1 MW gyrotron with the one-step depressed collector. *International Journal of Infrared and Millimeter Waves*, 18(11):2129–2136, 1997.

[GLP+11] G. Giruzzi, M. Lennholm, A. Parkin, G. Aiello, M. Bellinger, J. Bird, F. Bouquey, H. Braune, A. Bruschi, P. Butcher, R. Clay, E. de la Luna, G. Denisov, T. Edlington, J. Fanthome, D. Farina, J. Farthing, L. Figini, S. Garavaglia, J. Garcia, M. Gardener, T. Gerbaud, G. Granucci, J. Hay, M. Henderson, S. Hotchin, V.N. Ilyin, M. Jennison, W. Kasparek, P. Khilar, N. Kirneva, D. Kislov, S. Knipe, A. Kuyanov, X. Litaudon, A.G. Litvak, A. Moro, S. Nowak, V. Parail, B. Plaum,

G. Saibene, C. Sozzi, P. Späh, D. Strauss, E. Trukhina, A. Vaccaro, A. Vagdama, V. Vdovin, and JET EFDA Contributors. Objectives, physics requirements and conceptual design of an ECRH system for JET. *Nuclear Fusion*, 51(6):063033, 2011.

[GSA+14] G. Gantenbein, A. Samartsev, G. Aiello, G. Dammertz, J. Jelonnek, M. Losert, A. Schlaich, T. Scherer, D. Strauss, M. Thumm, and D. Wagner. First operation of a step-frequency tunable 1-MW gyrotron with a diamond brewster angle output window. *IEEE Transactions on Electron Devices*, 61(6):1806–1811, 2014.

[HB75] E. O. Hammerstad and F. Bekkadal. *Microstrip handbook*. Electronics Research Laboratory, University of Trondheim, Norwegian Institute of Technology, 1975.

[HDG+09] R. Hemsworth, H. Decamps, J. Graceffa, B. Schunke, M. Tanaka, M. Dremel, A. Tanga, H.P.L. De Esch, F. Geli, J. Milnes, T. Inoue, D. Marcuzzi, P. Sonato, and P. Zaccaria. Status of the ITER heating neutral beam system. *Nuclear Fusion*, 49(4):045006, 2009.

[HOA+05] R Hiwatari, K Okano, Y Asaoka, K Shinya, and Y Ogawa. Demonstration tokamak fusion power plant for early realization of net electric power generation. *Nuclear Fusion*, 45(2):96, 2005.

[IAE00] *Energy for Tomorrow*. International Atomic Energy Agency (IAEA), 2000.

[IFMC10] R. L. Ives, L. R. Falce, G. Miram, and G. Collins. Controlled-porosity cathodes for high-current-density applications. *IEEE Transactions on Plasma Science*, 38(6):1345–1353, 2010.

[IFSW05] R. L. Ives, L. R. Falce, S. Schwartzkopf, and R. Witherspoon. Controlled porosity cathodes from sintered tungsten wires. *IEEE Transactions on Electron Devices*, 52(12):2800–2805, 2005.

[Igo15] V. Igochine, editor. *Active Control of Magneto-hydrodynamic Instabilities in Hot Plasmas.* Springer Berlin Heidelberg, 2015.

[IKF+17] S. Illy, P. C. Kalaria, J. Franck, K. A. Avramidis, G. Gantenbein, I. Gr. Pagonakis, M. Thumm, and J. Jelonnek. Investigation on misalignment tolerances of 240-GHz demo gyrotrons. In *18th International Vacuum Electronics Conference. London, United Kingdom*, 2017.

[ILN+01] R. L. Ives, W. Lawson, J. M. Neilson, M. Read, P. L. Colestock, and S. Kelley. Development of a 10 MW, 91 GHz gyroklystron for accelerator applications. In *AIP Conference Proceedings*, volume 569, pages 663–678, 2001.

[IOK+16] R. Ikeda, Y. Oda, T. Kobayashi, K. Kajiwara, M. Terakado, K. Takahashi, S. Moriyama, and K. Sakamoto. Multi-frequency, MW-power triode gyrotron having a uniform directional beam. *Journal of Infrared, Millimeter, and Terahertz Waves*, 2016.

[IPDR04] S. Illy, B. Piosczyk, I. Danilov, and S. Raff. Design studies of the collector sweeping system for the 2 MW 170 GHfz coaxial gyrotron for ITER. In *the Joint 29th International Conference on Infrared and Millimeter Waves 2004 and 12th International Conference on Terahertz Electronics*, 2004.

[ISN+98] T. Idehara, K. Shibutani, H. Nojima, M. Pereyaslavets, K. Yoshida, I. Ogawa, and T. Tatsukawa. Study of electron beam misalignment in a submillimeter wave gyrotron. *International Journal of Infrared and Millimeter Waves*, 19(10):1303–1316, 1998.

[IWK+95] C. T. Iatrou, D. R. Whaley, S. Kern, M. Thumm, M. Q. Tran, A. Möbius, H. U. Nickel, P. Norajitra, A. Wien, T. M. Tran, G. Bon Mardion, M. Pain, and G. Tonon. Feasibility study of the EU home team on a 170 GHz 1 MW CW gyrotron for ECH on ITER. *International Journal of Infrared and Millimeter Waves*, 16(6):1129–1158, 1995.

[JAA+15] J. Jelonnek, G. Aiello, K. Avramidis, J. Franck, G. Gantenbein, G. Grossetti, K. Hesch, S. Illy, J. Jin, P. Kalaria, I. G. Pagonakis, T. Rzesnicki, S. Ruess, A. Samartsev, T. Scherer, D. Strauss, C. Wu, M. Thumm, S. Alberti, F. Braunmueller, J. Genoud, J. P. Hogge, C. Schlatter, Trach-Minh Tran, Minh Quang Tran, I. Chelis, J. L. Vomvoridis, Z. C. Ioannidis, G. Latsas, I. G. Tigelis, A. Bruschi, M. Lontano, V. Hermann, F. Legrand, Y. Rozier, F. Albajar, T. Bonicelli, and F. Cismondi. From W7-X towards iter and beyond: Status and progress in eu fusion gyrotron developments. In *16th International Vacuum Electronics Conference (IVEC)*, pages 1–2, April 2015.

[JAF+14] J. Jelonnek, K. Avramidis, J. Franck, G. Dammertz, G. Gantenbein, K. Hesch, S. Illy, J. Jin, A. Malygin, I. Gr. Pagonakis, et al. KIT contribution to the gyrotron development for nuclear fusion experiments in europe. In *8th German Microwave Conference (GeMiC)*, pages 1–4, 2014.

[Jel00] J. Jelonnek. *Untersuchung des Lastverhaltens von Gyrotrons*. VDI-Verlag, 2000.

[JFJ+13] J. Jin, J. Flamm, J. Jelonnek, S. Kern, I. Pagonakis, T. Rzesnicki, and M. Thumm. High efficiency quasi optical mode converter for a 1 MW $TE_{32,9}$ mode gyrotron. *IEEE Transactions on Plasma Science*, 41(10):2748–2753, 2013.

[JGJ+15] J. Jin, G. Gantenbein, J. Jelonnek, T. Rzesnicki, and M. Thumm. Development of mode conversion waveguides at KIT. *EPJ Web of Conferences*, 87:04003, 2015.

[JGS99] J. Jelonnek, A. Grudiev, and K. Schunemann. Rigorous computation of time-dependent electromagnetic fields in gyrotron cavities excited by internal sources. *IEEE Transactions on Plasma Science*, 27(2):374–383, 1999.

[JK10] R. Jain and M. V. Kartikeyan. Design of a 60 GHz, 100 kW CW gyrotron for plasma diagnostics: GDS-V.01 simulations. *Progress In Electromagnetics Research B*, 22:379–399, 2010.

[JTGJ16] J. Jin, M. Thumm, G. Gantenbein, and J. Jelonnek. A numerical synthesis method for hybrid-type high-power gyrotron launchers. *IEEE Transactions on Microwave Theory and Techniques*, pages 1–8, 2016.

[JTP+08] J. Jin, M. Thumm, B. Piosczyk, J. Flamm, O. Prinz, and T. Rzesnicki. Improved quasi-optical launcher for coaxial cavity ITER gyrotron. In *33rd International Conference on Infrared, Millimeter and Terahertz Waves (IRMMW-THz)*, 2008.

[JTP+09] J. Jin, M. Thumm, B. Piosczyk, S. Kern, J. Flamm, and T. Rzesnicki. Novel numerical method for the analysis and synthesis of the fields in highly oversized waveguide mode converters. *IEEE Transactions on Microwave Theory and Techniques*, 57(7):1661–1668, 2009.

[JTPR06] J. Jin, M. Thumm, B. Piosczyk, and T. Rzesnicki. Theoretical investigation of an advanced launcher for a 2-MW 170-GHz $TE_{34,19}$ coaxial cavity gyrotron. *IEEE Transactions on Microwave Theory and Techniques*, 54(3):1139–1145, 2006.

[KAF+16a] P. Kalaria, K. Avramidis, J. Franck, G. Gantenbein, S. Illy, I. Gr. Pagonakis, M. Thumm, and J. Jelonnek. A fast frequency step-tunable 236 GHz gyrotron design for demo. In *10th German Microwave Conference (GeMiC)*, pages 377–380, 2016.

[KAF+16b] P. Kalaria, K. Avramidis, J. Franck, G. Gantenbein, S. Illy, I. Gr. Pagonakis, M. Thumm, and J. Jelonnek. Investigation on mode eigenvalue limits for stable 236 GHz, 1 MW-class gyrotron operation. In *18th IEEE International Vacuum Electronics Conference (IVEC)*. Institute of Electrical and Electronics Engineers (IEEE), 2016.

[KAF+16c] P. Kalaria, K. Avramidis, J. Franck, G. Gantenbein, S. Illy, I. Gr. Pagonakis, M. Thumm, and J. Jelonnek. Systematic cavity design approach for a multi-frequency gyrotron for DEMO and study of its RF behavior. *Physics of Plasmas*, 23(9):092503, 2016.

[KBT10] M. V. Kartikeyan, E. Borie, and M. Thumm. *Gyrotrons: High-Power Microwave and Millimeter Wave Technology*. Springer, 2010.

[KDNI14] E. M. Khutoryan, O. Dumbrajs, G. S. Nusinovich, and T. Idehara. Theoretical study of the effect of electron beam misalignment on operation of the gyrotron FU IV a. *IEEE Transactions on Plasma Science*, 42(6):1586–1593, 2014.

[Ker96] S. Kern. *Numerische Simulation der Gyrotron- Wechselwirkung in koaxialen Resonatoren.* PhD thesis, Universität Karlsruhe (TH), 1996.

[KHA+12] S. Kern, J.-P. Hogge, S. Alberti, K. Avramides, G. Gantenbein, S. Illy, J. Jelonnek, J. Jin, F. Li, I. Gr. Pagonakis, B. Piosczyk, T. Rzesnicki, M. K. Thumm, I. Tigelis, and M. Q. Tran and. Experimental results and recent developments on the EU 2 MW 170 GHz coaxial cavity gyrotron for ITER. *EPJ Web of Conferences*, 32:04009, 2012.

[KIP+16] K. Kim, K. Im, H Park, Y. K. Oh, and Y. S. Kim. Recent fusion activities of korea toward demo. In *Technology for Fusion Energy*, 2016.

[KIS+12] T. Kobayashi, A. Isayama, M. Sawahata, S. Suzuki, M. Terakado, S. Hiranai, K. Wada, Y. Sato, J. Hinata, K. Yokokura, et al. Dual frequency gyrotron development for JT-60SA. In *37th International Conference on Infrared, Millimeter, and Terahertz Waves (IRMMW-THz), 2012*, pages 1–2. IEEE, 2012.

[KKO+13] K. Kim, H. C. Kim, S. Oh, Y. S. Lee, Yeom J. H., Im K., G. S. Lee, G. Neilson, C. Kessel, T. Brown, and P. Titus. A preliminary conceptual design study for korean fusion DEMO reactor. *Fusion Engineering and Design*, 88(6-8):488–491, 2013.

[KLTZ92] A. N. Kuftin, V. K. Lygin, Sh. E. Tsimiring, and V. E. Zapevalov. Numerical simulation and experimental study of magnetron-injection guns for powerful short-wave gyrotrons. *International Journal of Electronics*, 72(5-6):1145–1151, 1992.

[KM16] A. N. Kuftin and V. N. Manuilov. The electron-optical system of a gyrotron with an operating frequency of 263 GHz for spectroscopic research. *Radiophysics and Quantum Electronics*, 59(2):130–136, 2016.

[KNSA10] D. G. Kashyn, G. S. Nusinovich, O. V. Sinitsyn, and T. M. Antonsen. Single-mode excitation in high-power gyrotrons by controlling gun perveance. *IEEE Transactions on Plasma Science*, 38(6):1160–1167, jun 2010.

[KOK+11] K. Kajiwara, Y. Oda, A. Kasugai, K. Takahashi, and K. Sakamoto. Development of dual-frequency gyrotron with triode magnetron injection gun. *Applied Physics Express*, 4(12):126001, nov 2011.

[KSK+11] N. Kumar, U. Singh, A. Kumar, H. Khatun, and A. K. Sinha. On the design of a high-efficiency double-beam gyrotron. *IEEE Transactions on Plasma Science*, 39(9):1781–1785, 2011.

[KST+08] A. Kasugai, K. Sakamoto, K. Takahashi, K. Kajiwara, and N. Kobayashi. Steady-state operation of 170 GHz–1 MW gyrotron for ITER. *Nuclear Fusion*, 48(5):054009, 2008.

[KT80] K. E. Kreischer and R. J. Temkin. Linear theory of an electron cyclotron maser operating at the fundamental. *International Journal of Infrared and Millimeter Waves*, 1(2):195–223, 1980.

[KT87] K. E. Kreischer and R. J. Temkin. Single-mode operation of a high-power, step-tunable gyrotron. *Physical Review Letters*, 59(5):547–550, 1987.

[KYS+15] T. Kobayashi, S. Moriyama K. Yokokura, M. Sawahata, M. Terakado, S. Hiranai, K. Wada, Y. Sato, J. Hinata, K. Hoshino, A. Isamaya, Y. Oda, R. Ikeda, K. Takahashi, and

K. Sakamoto. Gyrotron development for high-power, long-pulse electron cyclotron heating and current drive at two frequencies in JT-60SA and its extension toward operation at three frequencies. *Nuclear Fusion*, 55(6):063008, 2015.

[LA90] B. Levush and T.M. Antonsen. Mode competition and control in higher-power gyrotron oscillators. *IEEE Transactions on Plasma Science*, 18(3):260–272, 1990.

[Law57] J. D. Lawson. Some criteria for a power producing thermonuclear reactor. *Proceedings of the Physical Society. Section B*, 70(1):6–10, 1957.

[LDM+12] A.G. Litvak, G.G. Denisov, V.E. Myasnikov, E.M. Tai, E.V. Sokolov, and V.I. Ilin. Recent development results in russia of megawatt power gyrotrons for plasma fusion installations. *EPJ Web of Conferences*, 32:04003, 2012.

[LHC+10] F. Li, W. He, A. W. Cross, C. R. Donaldson, L. Zhang, A. D. R. Phelps, and K. Ronald. Design and simulation of a 390 GHz seventh harmonic gyrotron using a large orbit electron beam. *Journal of Physics D: Applied Physics*, 43(15):155204, 2010.

[LN84] A. G. Luchinin and G. S. Nusinovich. An analytical theory for comparing the efficiency of gyrotrons with various electrodynamic systems. *International Journal of Electronics*, 57(6):827–834, 1984.

[LNY+13] Ying-hui Liu, Xin-jian Niu, Sheng Yu, Hong-fu Li, Jun Rao, Jun Zhou, Chaojun Lei, Hong Jiang, and Yi Yang. Design and experimental demonstration of a 94-ghz gyrotron with complex cavity operating at the second harmonic. *Journal of Fusion Energy*, 32(2):182–188, 2013.

[LS13] M. C. Lin and D. N. Smithe. Modeling a gyrotron cavity using a 3D CFDTD PIC method. In *16th International Vacuum Electronics Conference (IVEC)*, 2013.

[LSC⁺14] M. C. Lin, D. N. Smithe, E. Choi, K. R. Chu, W. C. Guss, and R. J. Temkin. Cold test of gyrotron cavity modes using a 3D CFDTD method. In *15th International Vacuum Electronics Conference (IVEC)*, 2014.

[LST11] A. Litvak, K. Sakamoto, and M. Thumm. Innovation on high-power long-pulse gyrotrons. *Plasma Physics and Controlled Fusion*, 53(12):124002, 2011.

[Mal16] A. Malygin. *Design and Experimental Investigation of a Second Harmonic 20 kW Class 28 GHz Gyrotron for Evaluation of New Emitter Technologies*. PhD thesis, Karlsruher Institute für Technologies, 2016.

[MCC⁺07] D. Maisonnier, D. Campbell, I. Cook, L. Di Pace, L. Giancarli, J. Hayward, A. Li Puma, M. Medrano, P. Norajitra, M. Roccella, P. Sardain, M.Q. Tran, and D. Ward. Power plant conceptual studies in europe. *Nuclear Fusion*, 47(11):1524–1532, 2007.

[Mor92] A. W. Morris. MHD instability control, disruptions, and error fields in tokamaks. *Plasma Physics and Controlled Fusion*, 34(13):1871–1879, 1992.

[MRR⁺16] M. M. Melnikova, A. G. Rozhnev, N. M. Ryskin, T. Saito, Y. Tatematsu, and Y. Yamaguchi. Modeling of a complex-cavity resonator for the 0.4 THz second-harmonic frequency-tunable gyrotron. In *Infrared, Millimeter, and Terahertz waves (IRMMW-THz), 2016 41st International Conference on*, 2016.

[MS16] V. N. Manuilov and V. E. Semenov. Ion compensation of the space charge in the gyrotron helical electron beams. In *41st International Conference on Infrared, Millimeter, and Terahertz waves (IRMMW-THz)*, 2016.

[MSBL98] D. B. McDermott, R.C. Statzman, A.J. Balkcum, and N.C. Luhmann. 94-GHz 25-kW CW low-voltage harmonic gyrotron. *IEEE Transactions on Plasma Science*, 26(3):402–408, 1998.

[MSMS04] V.N. Manuilov, D.A. Smirnov, S.A. Malygin, and F.-A. Soluyanova. Numerical simulation of the gyrotron collector systems with rotating magnetic field. In *Joint 29th International Conference on Infrared and Millimeter Waves and 12th International Conference on Terahertz Electronics*, 2004.

[MT93] A. Möbius and M. Thumm. *Gyrotron Oscillators - their principles and practice*, chapter Gyrotron output launchers and output tapers, pages 179–222. Taylor Francis, 1993.

[NDL95] G. S. Nusinovich, O. Dumbrajs, and B. Levush. Wave interaction in gyrotrons with off-axis electron beams. *Physics of Plasmas*, 2(12):4621–4630, 1995.

[Nei04] J. Neilson. Surf3d and LOT: computer codes for design and analysis of high-perfonnance QO launchers in gyrotrons. In *12th International Conference on Terahertz Electronics*, 2004.

[Nus81] G. S. Nusinovich. Mode interaction in gyrotrons. *International Journal of Electronics Theoretical and Experimental*, 51(4):457–474, 1981.

[Nus04] G. S. Nusinovich. *Introduction to the Physics of Gyrotrons*. Johns Hopkins University Press, 2004.

[NVA+08] G. S. Nusinovich, A. N. Vlasov, T. M. Antonsen, J. Lohr, B. G. Danly, and J.-P. Hogge. Excitation of parasitic modes in gyrotrons with fast voltage rise. *Physics of Plasmas*, 15(10):103101, 2008.

[OP09] Gerd Gantenbein Ying-hui Liu Manfred Thumm Dietmar Wagner Oliver Prinz, Andreas Arnold. Highly efficient quasi-optical mode converter for a multifrequency high-power gyrotron. *IEEE Transactions on Electron Devices*, 56(5):828–834, 2009.

[PADT07] H. O. Prinz, A. Arnold, G. Dammertz, and M. Thumm. Redesign and 3D-simulation of the quasi-optical mode converter for a $TE_{22,6}$ 118 GHz gyrotron. In *2007 IEEE Pulsed Power Plasma Science Conference*, 2007.

[PAI+14] I. Gr. Pagonakis, K. Avramidis, S. Illy, P. Kalaria, G. Gantenbein, and J. Jelonnek. Electron beam simulation in the overall gyrotron geometry. In *9th International Workshop on Strong Microwaves and Terahertz Waves: Sources and Applications*, 2014.

[PBB+09] J. Pamela, A. Bécoulet, D. Borba, J.-L. Boutard, L. Horton, and D. Maisonnier. Efficiency and availability driven R&D issues for DEMO. *Fusion Engineering and Design*, 84(2-6):194–204, 2009.

[PDD+04] B. Piosczyk, G. Dammertz, O. Dumbrajs, M.V. Kartikeyan, M.K. Thumm, and X. Yang. 165-GHz coaxial cavity gyrotron. *IEEE Transactions on Plasma Science*, 32(3):853–860, 2004.

[PEK+10] Burkhard Plaum, Volker Erckmann, Walter Kasparek, Heinrich Laqua, Carsten Lechte, and Michael Weisgerber. Design of a remote steering ECRH launcher for the stellerator wendelstein 7-x. In *35th International Conference on Infrared, Millimeter, and Terahertz Waves*, 2010.

[PIDT96] B. Piosczyk, C.T. Iatrou, G. Dammertz, and M. Thumm. Single-stage depressed collectors for gyrotrons. *IEEE Transactions on Plasma Science*, 24(3):579–585, 1996.

[Pio90] B. Piosczyk. Einfluss der raumladungskompensation auf den betrieb eines gyrotrons. Technical report, Institut für Technische Physik, Kernforschungzentrum Karlsruhe GmbH, 1990.

[PIT16] I. Gr. Pagonakis, S. Illy, and M. Thumm. Influence of emitter ring manufacturing tolerances on electron beam quality of high power gyrotrons. *Physics of Plasmas*, 23(8):083103, 2016.

[PTZ^{+}12] E. Poli, G. Tardini, H. Zohm, E. Fable, D. Farina, L. Figini, N.B. Marushchenko, and L. Porte. Electron-cyclotron-current-drive efficiency in DEMO plasmas. *Nuclear Fusion*, 53(1):013011, 2012.

[PV04] I. Gr. Pagonakis and J. L. Vomvoridis. The self-consistent 3D trajectory electrostatic code ARIADNE for gyrotron beam tunnel simulation. In *29th International Conference on Infrared and Millimeter Waves and 12th International Conference on Terahertz Electronics*, 2004.

[PWI^{+}16] I. Gr. Pagonakis, C. Wu, S. Illy, G. Gantenbein, and J. Jelonnek. Multistage depressed collector design approach for gyrotron. In *17th International Vacuum Electronics Conference (IVEC)*, 2016.

[PWIJ16] I. Gr. Pagonakis, C. Wu, S. Illy, and J. Jelonnek. Multistage depressed collector conceptual design for thin magnetically confined electron beams. *Physics of Plasmas*, 23(4):043114, 2016.

[RAA+17] T. Rzesnicki, F. Albajar, S. Alberti, K.A. Avramidis, W. Bin, T. Bonicelli, F. Braunmueller, A. Bruschi, J. Chelis, P.-E. Frigot, G. Gantenbein, V. Hermann, J.-P. Hoggc, S. Illy, Z.C. Ioannidis, J. Jin, J. Jelonnek, W. Kasparek, G.P. Latsas, C. Lechte, M. Lontano, T. Kobarg, I.G. Pagonakis, Y. Rozier, C. Schlatter, M. Schmid, I.G. Tigelis, M. Thumm, M.Q. Tran, J.L. Vomvoridis, and A. Zisis. Experimental verification of the European 1 MW, 170 GHz industrial CW prototype gyrotron for ITER. *Fusion Engineering and Design*, 2017.

[Rom12] F. Romanelli. Fusion electricity, a roadmap to the realization of fusion energy. Technical report, European Fusion Development Agreement (EFDA), 2012.

[RPG+16] S. Ruess, I. Gr. Pagonakis, G. Gantenbein, S. Illy, T. Kobarg, T. Rzesnicki, M. Thumm, J. Weggen, and J. Jelonnek. An inverse magnetron injection gun for the KIT 2-MW coaxial-cavity gyrotron. *IEEETransactions on Electron Devices*, 63(5):2104–2109, 2016.

[RPK+10] T. Rzesnicki, B. Piosczyk, S. Kern, S. Illy, J. Jin, A. Samartsev, A. Schlaich, and M. Thumm. 2.2-MW record power of the 170-GHz european preprototype coaxial-cavity gyrotron for ITER. *IEEE Transactions on Plasma Science*, 38(6):1141–1149, 2010.

[RPR+15] S. Ruess, I. Gr Pagonakis, T. Rzesnicki, G. Gantenbein, S. Illy, M. Thumm, and J. Jelonnek. A design proposal for an optimized inverse magnetron injection gun for the KIT 2 MW/170 GHz modular coaxial cavity gyrotron. In *International Vacuum Electronics Conference (IVEC)*, pages 1–2, 2015.

[RRP+16] S. Ruess, T. Rzesnicki, I. Gr. Pagonakis, T. Kobarg, M. Fuchs, S. Illy, G. Gantenbein, M. Thumm, and J. Jelonnek. Experimental results and outlook of the 2 MW 170 GHz coaxial-cavity gyrotron towards long pulse operation. In *German Microwave Conference (GeMiC)*, pages 369–372, 2016.

[SALL92] G.P. Saraph, T.M. Antonsen, B. Levush, and G.I. Lin. Regions of stability of high-power gyrotron oscillators. *IEEE Transactions on Plasma Science*, 20(3):115–125, 1992.

[SANL93] G. P. Saraph, T. M. Antonsen, G. S. Nusinovich, and B. Levush. Nonlinear theory of stable, efficient operation of a gyrotron at cyclotron harmonics. *Physics of Fluids B: Plasma Physics*, 5(12):4473–4485, 1993.

[SBF+12] C. Sozzi, T. Bolzonella, D. Farina, L. Figini, M. Furukawa, G. Giruzzi, S. Ide, A. Isayama, G. Matsunaga, S. Moriyama, S. Nowak, and and. ECH and ECCD effects on NTMs stabilization by ECRF in JT-60sa tokamak. *EPJ Web of Conferences*, 32:02017, 2012.

[SBN+15] L. Savoldi, A. Bertinetti, G. F. Nallo, A. Zappatore, R. Zanino, F. Cau, F. Cismondi, and Y. Rozier. CFD analysis of mini-channel cooling for a gyrotron cavity. In *26th Symposium on Fusion Engineering (SOFE)*, 2015.

[SBN+16] L. Savoldi, A. Bertinetti, G. F. Nallo, A. Zappatore, R. Zanino, F. Cau, F. Cismondi, and Y. Rozier. CFD analysis of different cooling options for a gyrotron cavity. *IEEE Transactions on Plasma Science*, 44(12):3432–3438, 2016.

[Sch15] A. Schlaich. *Time-dependent spectrum analysis of high power gyrotrons*. PhD thesis, Karlsruher Institut für Technologie, 2015.

[SDG+13] A. Samartsev, G. Dammertz, G. Gantenbein, J. Jelonnek, S. Illy, and M. Thumm. Influence of annular beam displacement on the performance of a high-power gyrotron. *IEEE Transactions on Plasma Science*, 41(4):872–878, 2013.

[SGH+91] A. Singh, V. L. Granatstein, G. Hazel, G. Saraph, J. M. Cooperstein, and T. Hargreaves. A depressed collector system for a quasi-optical gyrotron with precisely controlled magnetic flux lines. *International Journal of Infrared and Millimeter Waves*, 12(4):323–334, 1991.

[Sho93] O. Shojiro. *Mechanical properties of metallic composites*. CRC Press, 1993.

[SJ81] R. S. Symons and H. R. Jory. Cyclotron resonance devices. *Advances in Electronics and Electron Physics*, 55:1–75, 1981.

[SJB+98] D. F. H. Start, J. Jacquinot, V. Bergeaud, V. P. Bhatnagar, G. A. Cottrell, S. Clement, L-G. Eriksson, A. Fasoli, A. Gondhalekar, C. Gormezano, G. Grosshoeg, K. Guenther, P. Harbour, L. D. Horton, A. Howman, H. Jackel, O. N. Jarvis, K. D. Lawson, C. Lowry, M. Mantsinen, F. B. Marcus, R. Monk, E. Righi, F. G. Rimini, G. J. Sadler, G. R. Saibene, R. Sartori, B. Schunke, S. Sharapov, A. C. C. Sips, M. Stamp, and P. Belle. D-T fusion with ion cyclotron resonance heating in the JET tokamak. *Physical Review Letters*, 80(21):4681–4684, 1998.

[SKH+14] A. Sawant, J. H. Kim, Y. Hong, J. So, and E. Choi. Comparative study on PIC simulations of the gyrotron cavity. In *39th International Conference on Infrared, Millimeter, and Terahertz waves (IRMMW-THz)*, 2014.

[SKK+09] K. Sakamoto, A. Kasugai, K. Kajiwara, K. Takahashi, Y. Oda, K. Hayashi, and N. Kobayashi. Progress of high power 170 GHz gyrotron in JAEA. *Nuclear Fusion*, 49(9):095019, 2009.

[SKL+14] A. Sawant, S. G. Kim, M.-C. Lin, J. H. Kim, Y. Hong, J. So, and E. Choi. Validation of 3-D time domain particle-in-cell simulations for cold testing a W-band gyrotron cavity. *IEEE Transactions on Plasma Science*, 42(12):3989–3995, 2014.

[SKO+12] K. Sakamoto, K. Kajiwara, Y. Oda, K. Takahashi, K. Hayashi, and N. Kobayashi. Progress of high power multi-frequency gyrotron development. In *37th International Conference on Infrared, Millimeter, and Terahertz Waves (IRMMW-THz)*, 2012.

[SKO+13] K. Sakamoto, K. Kajiwara, Y. Oda, K. Hayashi, K. Takahashi, T. Kobayashi, and S. Moriyama. Status of high power gyrotron development in JAEA. In *14th International Vacuum Electronics Conference (IVEC)*, 2013.

[SKT+07] K. Sakamoto, A. Kasugai, K. Takahashi, R. Minami, N. Kobayashi, and K. Kajiwara. Achievement of robust high-efficiency 1 MW oscillation in the hard-self-excitation region by a 170 GHz continuous-wave gyrotron. *Nature Physics*, 3(6):411–414, 2007.

[SNAV08] O. V. Sinitsyn, G. S. Nusinovich, T. M. Antonsen, and A. N. Vlasov. Startup scenarios in MW-class gyrotrons with diode and triode-type electron guns. In *33th International Conference on Infrared, Millimeter, and Terahertz waves (IRMMW-THz)*, 2008.

[SSH+01] H. Shoyama, K. Sakamoto, K. Hayashi, A. Kasugai, M. Tsuneoka, K. Takahashi, Y. Ikeda, T. Kariya, Y. Mitsunaka, and T. Imai. High-efficiency oscillation of 170 GHz high-power gyrotron at $TE_{31,8}$ mode using depressed collector. *Japanese Journal of Applied Physics*, 40:906–908, 2001.

[STK+06a] K. Sakamoto, K. Takahashi, A. Kasugai, R. Minami, N. Kobayashi, S. Nishio, M. Sato, and K. Tobita. Conceptual study of ECH/ECCD system for fusion DEMO plant. *Fusion Engineering and Design*, 81(8-14):1263–1270, 2006.

[STK+06b] K. Sakamoto, K. Takahashi, A. Kasugai, R. Minami, N. Kobayashi, S. Nishio, M. Sato, and K. Tobita. Conceptual study of ECH/ECCD system for fusion DEMO plant. *Fusion Engineering and Design*, 81(8-14):1263–1270, 2006.

[STK+06c] K. Sakamoto, K. Takahashi, A. Kasugai, R. Minami, N. Kobayashi, S. Nishio, M. Sato, and K. Tobita. Conceptual study of ECH/ECCD system for fusion DEMO plant. *Fusion Engineering and Design*, 81(8-14):1263–1270, 2006.

[SWRG97] A. Singh, D.S. Weile, S. Rajapatirana, and V.L. Granatstein. Integrated design of depressed collectors for gyrotrons. *IEEE Transactions on Plasma Science*, 25(3):480–491, 1997.

[TAB+01] M. Thumm, A. Arnold, E. Borie, O. Braz, G. Dammertz, O. Dumbrajs, K. Koppenburg, M. Kuntze, G. Michel, and B. Piosczyk. Frequency step-tunable (114–170 GHz) megawatt gyrotrons for plasma physics applications. *Fusion Engineering and Design*, 53(1):407–421, 2001.

[TAH+01] M. Thumm, A. Arnold, R. Heidinger, M. Rohde, R. Schwab, and R. Spoerl. Status report on CVD-diamond window development for high power ECRH. *Fusion Engineering and Design*, 53(1-4):517–524, 2001.

[TBDG10] L. Tsang, H. Braunisch, R. Ding, and X. Gu. Random rough surface effects on wave propagation in interconnects. *IEEE Transactions on Advanced Packaging*, 33(4):839–856, 2010.

[TFK+15] M. Thumm, J. Franck, P.C. Kalaria, K.A. Avramidis, G. Gantenbein, S. Illy, I.G. Pagonakis, M. Schmid, C. Wu, J. Zhang, and J. Jelonnek. Towards a 0.24 THz, 1 to 2 MW class gyrotron for demo. *International Journal of Terahertz Science and Technology*, 8(3), 2015.

[Thu03] M. Thumm. Progress in gyrotron development. *Fusion Engineering and Design*, 66-68:69–90, 2003.

[Thu11] M. Thumm. Progress on gyrotrons for iter and future thermonuclear fusion reactors. *IEEE Transactions on Plasma Science*, 39(4):971–979, 2011.

[Thu16] M. Thumm. *State-of-the-Art of High Power Gyro-Devices and Free Electron Masers. Update 2015 (KIT Scientific Reports ; 7717).* KIT Scientific Publishing, 2016.

[TSG+13] D. S. Tax, O. V. Sinitsyn, W. C. Guss, G. S. Nusinovich, M. A. Shapiro, and R. J. Temkin. Experimental study of the start-up scenario of a 1.5-MW, 110-GHz gyrotron. *IEEE Transactions on Plasma Science*, 41(4):862–871, 2013.

[Tsi01] Sh.E. Tsimring. Gyrotron electron beams: Velocity and energy spread and beam instabilities. *International Journal of Infrared and Millimeter Waves*, 22(10):1433–1468, 2001.

[TWL+15] Y. Tang, X. Wang, F. Liu, L. Zhang, W. Wei, H. Xu, W. Xu, D. Wu, and J. Feng. Conceptual design of CFETR electron cyclotron wave system. *Fusion Engineering and Design*, 94:48–53, 2015.

[VAC+08] A. N. Vlasov, T. M. Antonsen, I. A. Chernyavskiy, G. S. Nusinovich, and B. Levush. MAGY simulations of mode interaction in a coaxial gyrotron. *IEEE Transactions on Plasma Science*, 36(3):606–619, 2008.

[VZO+69] S. N. Vlasov, G. M. Zhislin, I. M. Orlova, M. I. Petelin, and G. G. Rogacheva. Irregular waveguides as open resonators. *Radiophysics and Quantum Electronics*, 12(8):972–978, 1969.

[VZP75] S. N. Vlasov, L. I. Zagriadskaia, and M. I. Petelin. Transformation of a whispering gallery mode, propagating in a circular waveguide, into a beam of waves. *Radio Engineering and Electronic Physics*, 20:2026–2030, 1975.

[WAA+15] R. Wenninger, F. Arbeiter, J. Aubert, L. Aho-Mantila, R. Albanese, R. Ambrosino, C. Angioni, J.-F. Artaud, M. Bernert, E. Fable, A. Fasoli, G. Federici, J. Garcia, G. Giruzzi, F. Jenko, P. Maget, M. Mattei, F. Maviglia, E. Poli, G. Ramogida, C. Reux, M. Schneider, B. Sieglin, F. Villone, M. Wischmeier, and H. Zohm. Advances in the physics basis for the european DEMO design. *Nuclear Fusion*, 55(6):063003, 2015.

[WDQ+13] B. Wan, S. Ding, J. Qian, G. Li, B. Xiao, and G. Xu. Physics of CFETR. In *25th Symposium on Fusion Engineering (SOFE)*, 2013.

[WNT95] L.M. Waganer, F. Najmabadi, and M.S. Tillack. What must DEMO do? In *16th International Symposium on Fusion Engineering*, 1995.

[WPG+17] C. Wu, I. Gr. Pagonakis, G. Gantenbein, S. Illy, M. Thumm, and J. Jelonnek. Conceptual designs of e × b multistage depressed collectors for gyrotrons. *Physics of Plasmas*, 24(4):043102, 2017.

[WPI$^+$16] C. Wu, I. Pagonakis, S. Illy, M. Thumm, G. Gantenbein, and J. Jelonnek. Preliminary studies on multistage depressed collectors for fusion gyrotrons. In *10th German Microwave Conference (GeMiC)*, 2016.

[WSK$^+$16] D. Wagner, J. Stober, M. Kircher, F. Leuterer, F. Monaco, M. Munich, M. Schubert, H. Zohm, G. Gantenbein, J. Jelonnek, M. Thumm, A. Meier, T. Scherer, D. Strauss, W. Kasparek, C. Lechte, B. Plaum, A. Zach, A. G. Litvak, G. G. Denisov, A. Chirkov, L. G. Popov, V. O. Nichiporenko, V. E. Myasnikov, E. M. Tai, E. A. Solyanova, and S. A. Malygin. Extension of the multi-frequency ECRH system at ASDEX upgrade. In *International Conference on Infrared, Millimeter, and Terahertz waves (IRMMW-THz)*, 2016.

[WTA$^+$95] D. R. Whaley, M. Q. Tran, S. Alberti, T. M. Tran, Jr. T. M. Antonsen, and C. Tran. Startup methods for single-mode gyrotron operation. *Physical Review Letters*, 75(7):1304–1307, 1995.

[WTTA94] D. R. Whaley, M. Q. Tran, T. M. Tran, and T. M. Antonsen. Mode competition and startup in cylindrical cavity gyrotrons using high-order operating modes. *IEEE Transactions on Plasma Science*, 22(5):850–860, 1994.

[XCYH14] N. Xinjian, L. Chaojun, L. Yinghui, and L. Hongfu. Study of the effect of the electron beam parameters on a 0.4 THz second harmonic gyrotron. *IEEE Transactions on Electron Devices*, 61(9):3324–3328, 2014.

[Xiz84] C. Xizhang. Experimental study of velocity spread in a cycloiding electron beam. *Journal of Electronics (China)*, 1(2):121–131, 1984.

[YST+17] Y. Yamaguchi, T. Saito, Y. Tatematsu, M. Fukunari, T. Hirobe, R. Shinbayashi, S. Tanaka, S. Kubo, T. Shimozuma, K. Tanaka, and M. Nishiura. Development of 303 ghz high power pulsed gyrotron. In *6th International Workshop on Far-Infrared Technologies (IW-FIRT 2017)*, Japan, 2017.

[ZGD+16] I. Zotova, N. Ginzburg, G. Denisov, R. Rozental, and A. Sergeev. Frequency locking and stabilization regimes in high-power gyrotrons with low-Q resonators. *Radiophysics & Quantum Electronics*, 58(9), 2016.

[Zha86] S. C. Zhang. On the gyrotron dispersion equation for a general mode. *International Journal of Electronics*, 60(2):297–299, 1986.

[Zha99] S. C. Zhang. Comparative study of coaxial-cavity and cylindrical-cavity gyrotron oscillators with electron-beam misalignment taken into account. In *Asia Pacific Microwave Conference*, volume 3, pages 833–835, 1999.

[Zha16] J. Zhang. *Influence of Emitter Surface Roughness and Emission Inhomogeneity on Efficiency and Stability of High Power Fusion Gyrotrons*. PhD thesis, Karlsruhe Institute of Technology, Germany, 2016.

[ZIP+15] J. Zhang, S. Illy, I. Gr Pagonakis, K. A Avramidis, M. Thumm, and J. Jelonnek. Influence of emitter surface roughness on high power fusion gyrotron operation. *Nuclear Fusion*, 56(2):026002, 2015.

[ZIP+17] J. Zhang, S. Illy, I. Gr. Pagonakis, T. Rzesnicki, K. A. Avramidis, A. Malygin, S. Ruess, A. Samartsev, G. Dammertz, B. Piosczyk, G. Gantenbein, M. Thumm, and J. Jelonnek.

Evaluation and influence of gyrotron cathode emission inhomogeneity. *IEEE Transactions on Electron Devices*, 2017.

[ZIPJ13] J. Zhang, S. Illy, I. Gr. Pagonakis, and J. Jelonnek. Preliminary study on the effects of emitter surface roughness on gyrotron electron beam quality. In *38th International Conference on Infrared, Millimeter, and Terahertz Waves (IRMMW-THz)*, 2013.

[ZSS07] G. I. Zaginaylov, V. I. Shcherbinin, and K. Schuenemann. Linear theory of electron cyclotron instability of electromagnetic waves in a magnetoactive plasma waveguide. *Plasma Physics Reports*, 33(8):637–645, 2007.

[ZT05] H. Zohm and M. Thumm. On the use of step-tuneable gyrotrons in ITER. *Journal of Physics: Conference Series*, 25:274–282, 2005.

[ZY17] A. Qixiang Zhao and B. Sheng Yu. The nonlinear designs and experiments on a 0.42-THz second harmonic gyrotron with complex cavity. *IEEE Transactions on Electron Devices*, 64(2):564–570, 2017.

[ZZM06] N. A. Zavolsky, V. E. Zapevalov, and M. A. Moiseev. Influence of the energy and velocity spread in the electron beam on the starting conditions and efficiency of a gyrotron. *Radiophysics and Quantum Electronics*, 49(2):108–119, 2006.

[ZZMS11] N. A. Zavol'skiy, V. E. Zapevalov, M. A. Moiseev, and A. S. Sedov. Influence of the axial misalignment of the electron beam and the cavity on the gyrotron parameters. *Radiophysics and Quantum Electronics*, 54(6):402–408, 2011.

List of Figures

List of Tables

Own publications as sole or principal author

1. **P. Kalaria**, K. A. Avramidis, J. Franck, G. Gantenbein , S. Illy, I. Gr. Pagonakis, M. Thumm, J. Jelonnek, "*RF behavior and launcher design for a fast frequency step-tunable 236 GHz gyrotron for DEMO*," Journal of RF-Engineering and Telecommunications - Frequenz, Vol. 71, No. 3-4, pp. 161-171, 2016. (DOI: 10.1515/freq-2016-0212)

2. **P. Kalaria**, K. A. Avramidis, J. Franck, G. Gantenbein , S. Illy, I. Gr. Pagonakis, M. Thumm, J. Jelonnek, "*Systematic cavity design approach for a multi-frequency gyrotron for DEMO and study of its RF behavior*," Physics of Plasmas, Vol. 23, No. 9, 2016. (DOI: 10.1063/1.4962238)

3. **P. Kalaria**, K. A. Avramidis, J. Franck, G. Gantenbein , S. Illy, I. Gr. Pagonakis, M . Thumm, and J. Jelonnek, "*A 236 GHz Hollow-Cavity Gyrotron Design for DEMO: Current Status and Operation Limits*," 6th KIT fusion Ph.D. seminar, 8 - 9 June 2016, Leinsweiler, Germany.

4. **P. Kalaria**, K. A. Avramidis, J. Franck, G. Gantenbein , S. Illy, I. Gr. Pagonakis, M . Thumm, and J. Jelonnek, "*Investigation on Mode Eigenvalue Limits for Stable 236 GHz,1 MW-Class Gyrotron Operation*," 17th IEEE International Vacuum Electronics Conference (IVEC-2016), 19 - 21 April 2016, Monterey CA, USA. (DOI: 10.1109/IVEC.2016.7561978)

5. **P. Kalaria**, K. A. Avramidis, J. Franck, G. Gantenbein , S. Illy, I. Gr. Pagonakis, M . Thumm, and J. Jelonnek, "*A Fast Frequency Step-tunable 236 GHz Gyrotron Design for DEMO*," 10th German Microwave Conference 2016 (GeMiC-2016), 14 - 16 March 2016, Bochum, Germany. (DOI: 10.1109/ GEMIC.2016.7461635)

6. **P. Kalaria**, K. A. Avramidis, J. Franck, G. Gantenbein , S. Illy, I. Gr. Pagonakis, M . Thumm, and J. Jelonnek, "*Conceptual Design Studies on 236 GHz Conventional Cavity Gyrotron for DEMO*," 27th Joint German-Russian Workshop on ECRH and Gytotrons, 29 June - 4 July 2015, Greifswald, Germany.

7. **P. Kalaria**, K. A. Avramidis, J. Franck, G. Gantenbein , S. Illy, I. Gr. Pagonakis, M . Thumm, and J. Jelonnek, "*High Frequency Conventional Cavity Gyrotron Design for DEMO*," 5th KIT fusion Ph.D. seminar, 16 - 17 June 2015, St. Martin, Germany.

8. **P. Kalaria**, K. A. Avramidis, J. Franck, G. Gantenbein , S. Illy, I. Gr. Pagonakis, M . Thumm, and J. Jelonnek, "*Multi-frequency Operation of DEMO Gyrotron with Realistic Electron Beam Parameters*," 16th IEEE International Vacuum Electronics Conference (IVEC-2015), 27 - 29 April 2015, Beijing, China. (Selected for best student paper contest) (DOI:10.1109/IVEC.2015.7223769)

9. **P. Kalaria**, K. A. Avramidis, J. Franck, G. Gantenbein , S. Illy, I. Gr. Pagonakis, M . Thumm, and J. Jelonnek, "*Interaction Circuit Design and RF Behavior of a 236 GHz Gyrotron for DEMO*," German Microwave Conference 2015 (GEMIC-2015), 16 - 18 March 2014, Nuremberg, Germany. (DOI: 10.1109/ GEMIC.2015.7107804)

10. **P. Kalaria**, K. A. Avramidis, J. Franck, G. Gantenbein , S. Illy, I. Gr. Pagonakis, M . Thumm, and J. Jelonnek, "*Conventional Cylindrical-Cavity Gyrotron Design For DEMO*," 4th ITG International Vacuum

Electronics Workshop 2014, 13 - 14 Oct. 2014, Bad honnef, Germany.

11. **P. Kalaria**, M.V. Kartikeyan, M. Thumm, "*Design of 170 GHz, 1.5 MW Conventional Cavity Gyrotron for Plasma Heating*," IEEE Trans. for Plasma Science, Vol. 42, No. 6, pp. 1522 - 1528, June 2014. (DOI: 10.1109/ TPS.2014.2305251)

12. **P. Kalaria**, A. Sharma, A. Sawant, J. Malik, S. L. Rao, M. V. Kartikeyan, M. Thumm "*Mode Purity Estimation of the Gyrotron RF beam*," Infrared, Millimeter, and Terahertz Waves (IRMMW-THz) Conference, 1 - 6 September 2013, Mainz, Germany. (DOI: 10.1109/ IRMMW-THz.2013.6665414)

13. **P. Kalaria**, M.V. Kartikeyan, "*Design Studies of Magnetron Injection Gun and Magnetic Guidance System for a 170 GHz, 500 KW CW Gyrotron*," National Conference on Vacuum Electronic Devices and Applications, 21-24 September 2012, Pilani, India.

14. **P. Kalaria**, M.V. Kartikeyan, "*Design Studies of Quasi-optical Launcher for a 170 GHz, 1.5 MW CW Gyrotron for ITER Project*," National Conference on Vacuum Electronic Devices and Applications, 21 - 24 September 2012, Pilani, India.

15. **P. Kalaria**, M.V. Kartikeyan, M. Thumm, "*Output System Design for 170 GHz, 0.5 MW Gyrotron for ECRH Application*," 13th IEEE International Vacuum Electronics Conference (IVEC-2012), 24 - 26 April 2012, Monterey, California. (DOI:10.1109/IVEC.2012.6262249)

16. **P. Kalaria**, M.V. Kartikeyan, M. Thumm, "*Design Studies of Quasi-optical Launcher of a 170 GHz, CW Gyrotron for ECRH Application*," International Conference on Microwaves, Antenna, propagation and Remote Sensing (ICMARS-2011), 7 - 10 December 2011, Jodhpur, India.

17. **P. Kalaria**, M.V. Kartikeyan, M. Thumm, "*Design Considerations of a 170 GHz, 0.5 MW, CW Gyrotron* ," International Conference on Microwaves, Antenna, propagation and Remote Sensing (ICMARS-2011), 7 - 10 December 2011, Jodhpur, India.

18. **P. Kalaria**, M.V. Kartikeyan, "*Modified CPW Fed Band-Notched Ultra-Wideband Antenna*," IEEE Applied Electromagnetics Conference (AEMC 2011), 18 - 22 December 2011, Kolkata, India.(Third Best Paper Award) (DOI:10.1109/AEMC.2011.6256777)

Further journal contributions

19. J. Jelonnek, G. Gantenbein, K. Avramidis, J. Franck, S. Illy, Z. Ioannidis, J. Jin, **P. Kalaria**, I. Pagonakis, S. Ruess and T. Rzesnicki, "*Gyrotron Forschung und Entwicklung am KIT*," Vakuum in Forschung und Praxis, Vol. 28, No. 6, pp.21-27, 2016. (DOI: 10.1002/vipr.201600631)

20. M . Thumm, J. Franck, **P. Kalaria**, K. A. Avramidis, G. Gantenbein , S. Illy, I. Gr. Pagonakis, C. Wu, J. Zhang, J. Jelonnek, "*Towards a 0.24 THz 1 to 2 MW Class Gyrotron for DEMO*," International Journal of Terahertz Science and Technology, Vol.8, No.3, Sept. 2015. (DOI: 10.11906/TST.085-100.2015.09.09)

21. M. Schmid, J. Franck, **P. Kalaria**, K. A. Avramidis, G. Gantenbein, S. Illy, J. Jelonnek, I. Gr. Pagonakis, T. Rzesnicki, M. Thumm, "*Gyrotron Development at KIT: FULGOR Test Facility and Gyrotron Concepts for DEMO*," Fusion Engineering and Design, Vol. 96-97, pp. 589 - 592, April 2015. (DOI: 10.1016/j.fusengdes.2015.03.003)

22. J. Malik, **P. Kalaria**, M.V. Kartikeyan, "*Transient response of Dual Band-Notched Ultra-wideband Antenna*," International Journal of Microwave and Wireless Technologies, Vol. 7, No. 1, pp. 1 - 7, March 2014. (DOI: 10.1017/S1759078714000348)

23. S. L. Rao, A. Sharma, M. Kushwah, **P. Kalaria**, T. Kumar Sharma, V. Rathod, R. Shah, D. Mandge, and G. Joshi " *Electron Cyclotron*

Power Source System For ITER, American Nuclear and Fusion Science and Technology, Vol. 65, No. 1, pp. 129-144, January 2014. (DOI: 10.13182/FST13-642)

24. J. Malik, **P. Kalaria**, M.V. Kartikeyan, "*Complementary Sierpinski Gasket Fractal Antenna for Dual Band WiMAX / WLAN (3.5/5.8 GHz) Applications*," International Journal of Microwave and Wireless Technologies, Vol. 5, No. 04, pp. 1 - 7, March 2013. (DOI: 10.1017/S1759078713000123)

Acknowledgments

This thesis was conducted at the Institute for Pulsed Power and Microwave Technology (IHM), Karlsruhe Institute of Technology (KIT), Germany with the support from German Academic Exchange Service (DAAD). This work would not have been accomplished without the support that I received from many people. Hereby, I have the most profound joy in acknowledging their support.

First and foremost, I want to express my sincere and deepest thanks to Prof. Dr. Ing. John Jelonnek for giving me the opportunity to work in a stimulating environment. His continuous support and valuable feedback at every stage of my research work encouraged me to give my best. Apart from the technical knowledge, I learned many professional skills from him, which will surely help me to shape my academic career.

My big thanks to Prof. Dr. M. V. Kartikeyan for showing kind interest in my work and being a second reviewer of my thesis. I would also like to thanks him for his continuous support and encouragement.

I would like to express my deep gratitude to Prof. Dr. rer. nat. Dr. h.c. Manferd Thumm for the kind guidance and for being a continuous source of motivation. Throughout the course of this work, discussions with Prof. Thumm initiated many activities and helped me to keep up to date with the latest advancement in the field of vacuum electron devices.

Furthermore, I am really grateful to Dr. Konstantinos Avramidis. His valuable support and time to time guidance on technical issues was significant in making this work a success. He helped me in every challenges and showed me the right direction.

A big thanks goes to Dr.-Ing. Joachim Franck for his support during my doctoral research time. He is the best research partner cum friend with whom I shared many nice moments, besides frequent technical discussions. Additionally, I would like to thank Dr. Stefan Illy, Dr. Gerd Gantenbein, Dipl.-Ing. Martin Schmid, Dr. Ioannis Pagonakis, Dr. Jianbo Jin Dr. Tomasz Rzesnicki and Dr. Zisis Ioannidis for their continuous support and friendly research environment in the institute. I would also like to thank secretariats of the institute, Mrs. Martina Huber, Mrs. Manuela Wettstein and Mrs. Bianka Seitz for supporting me in every possible way.

I kindly thank all my former and current Ph.D. colleagues at the IHM-KIT for providing me homely environment and their continued help, particularly, Dr.-Ing. Andreas Schlaich, Dr.-Ing. Jianghua Zhang, Dr.-Ing. Anton Malygin, Sebastian Ruess, Chuanren Wu, Vasileios Ramopoulos, Dominik Neumaier, Alexander Marek and Martin Hochberg.

I am thankful to Prof. Dr. Olgerts Dumbrajs, Institute of Solid State Physics, University of Latvia, for providing me the necessary technical suggestions for the electron beam misalignment tolerance analysis. Additionally, I also want to thank Dr. Falk Braunmüller from the Swiss Plasma Center, École Polytechnique Fédérale de Lausanne (EPFL) for his support to benchmark the velocity spread implementation of KIT codes with TWANG and TWANG-PIC codes (section A.5). Parts of the simulations presented in this work have been carried out using the HELIOS supercomputer at IFERC-CSC. I am really thankful for this support.

This work has been carried out within the framework of the EUROfusion Consortium and has received funding from the Euratom research and training programme 2014-2018 under grant agreement No 633053. The views and opinions expressed herein do not necessarily reflect those of the European Commission.

Above all, I thank my parents, Mrs. Dharmishtha Kalaria and Mr. Chandulal Kalaria, for their never losing trust and confidence in me, even in my tough time and become sources of inspiration. Additionally, very special appreciation to my lovely sister, Aemi Kalaria, who always take care of my smile. My big thanks to my friend Dr.-Ing. Bhawna Verma for her care, support and patience. Very special thanks to Veenu Kamra, Nitish Tripathi and my perfect flatmates, Kumar Ghosh and Kamlesh Ghael.

Karlsruhe, May 2017
Parth Kalaria

KARLSRUHER FORSCHUNGSBERICHTE AUS DEM
INSTITUT FÜR HOCHLEISTUNGSIMPULS- UND MIKROWELLENTECHNIK
(ISSN 2192-2764)

Die Bände sind unter www.ksp.kit.edu als PDF frei verfügbar
oder als Druckausgabe bestellbar.

Band 1 **MATTHIAS BERINGER**
Design Studies towards a 4 MW 170 GHz Coaxial-Cavity Gyrotron. 2011
ISBN 978-3-86644-663-2

Band 2 **JENS FLAMM**
Diffraction and Scattering in Launchers of
Quasi-Optical Mode Converters for Gyrotrons. 2012
ISBN 978-3-86644-822-3

Band 3 **MATTIA DEL GIACCO**
Investigation of Fretting Wear of Cladding Materials in Liquid Lead. 2013
ISBN 978-3-86644-960-2

Band 4 **AMITAVO ROY CHOUDHURY**
Investigations of After Cavity Interaction in Gyrotrons
Including the Effect of Non-uniform Magnetic Field. 2013
ISBN 978-3-7315-0129-9

Band 5 **MICHAEL BETZ**
The CERN Resonant WISP Search (CROWS). 2014
ISBN 978-3-7315-0199-2

Band 6 **ANDREAS SCHLAICH**
Time-dependent spectrum analysis of high power gyrotrons. 2015
ISBN 978-3-7315-0375-0

Band 7 **DHIDIK PRASTIYANTO**
Temperature- and Time-Dependent Dielectric Measurements
and Modelling on Curing of Polymer Composites. 2016
ISBN 978-3-7315-0424-5

Band 8 **YIMING SUN**
Adaptive and Intelligent Temperature Control of Microwave
Heating Systems with Multiple Sources. 2016
ISBN 978-3-7315-0467-2

KARLSRUHER FORSCHUNGSBERICHTE AUS DEM INSTITUT FÜR HOCHLEISTUNGSIMPULS- UND MIKROWELLENTECHNIK

(ISSN 2192-2764)

Die Bände sind unter www.ksp.kit.edu als PDF frei verfügbar
oder als Druckausgabe bestellbar.

Band 9 **JIANGHUA ZHANG**
Influence of Emitter surface roughness and Emission inhomogeneity on Efficiency and stability of high power Fusion gyrotrons. 2016
ISBN 978-3-7315-0578-5

Band 10 **ANTON MALYGIN**
Design and Experimental Investigation of a Second Harmonic 20 kW Class 28 GHz Gyrotron for Evaluation of New Emitter Technologies. 2016
ISBN 978-3-7315-0584-6

Band 11 **JOACHIM FRANCK**
Systematic Study of Key Components for a Coaxial-Cavity Gyrotron for DEMO. 2017
ISBN 978-3-7315-0652-2

Band 12 **PARTH CHANDULAL KALARIA**
Feasibility and Operational Limits for a 236 GHz Hollow-Cavity Gyrotron for DEMO. 2017
ISBN 978-3-7315-0717-8